Beyond the Technology Traps

Competing Effectively in the New Banking Industry Requires More than Technology and "Checking-a-Box"

Mark P. Dangelo
with Rick Grant

Library of Congress Cataloging-in-Publication Data

Author Names: Dangelo, Mark P., and Grant, Rick

Title: Beyond the Technology Traps

Description: First Edition | Includes Bibliographical References

Identifier: ISBN 978-0-9989235-0-5 (hardcover) | ISBN 978-0-9989235-1-2 (softcover)

BISAC Subjects: Business | Economic | Technology | Banks and Banking | Consumer Behavior | Culture | Information Managel ment | Leadership | Management | Innovative Relevance | Outsourcing | New Business Enterprises

Hardcover ISBN: 978-0-9989235-0-5

Softcover ISBN: 978-0-9989235-1-2

eBook ISBN: 978-0-9989235-2-9

Kindle ISBN: 978-0-9989235-3-6

ePDF ISBN: 978-0-9989235-4-3

Printed in the United States of America, First U.S. Edition.
Cover Designs by Mark P. Dangelo.

"I sincerely believe that banking establishments are more dangerous than standing armies, and that the principle of spending money to be paid by posterity, under the name of funding, is but swindling futurity on a large scale."

—Thomas Jefferson

Contents

NOTES

"What if we don't change at all ...
and something magical just happens?"

Image ©Shutterstock, Inc.

Preface

Mark Twain stated, "***What gets us into trouble is not what we don't know. It's what we know for sure that just ain't so***." Today, this is where Financial Services and Banking Organizations (FSBO's) find themselves. Stated differently and more bluntly, a friend of ours, in reference to FSBO's said, "No one is really doing anything that is innovative unless they are 'forced' to by the market—it a fear of eminent death." Why? There are too many traps. Too much can go wrong and end a career.

There are many FSBO leaders who cling to axioms that have long since expired—unable or unwilling to see the vast changes taking place from consumers to technology to widespread commoditization all within one of the most regulated industries in the world. It should be noted right up front, this is a primary reason why we have shied away from the term "banks" in favor of FSBO, as to achieve the latter definition, and one which we believe will dominate traditional banking, will demand material changes to business models, methods of delivery, markets, personnel, and of course, technology.

Before the Great Recession of 2008, **we knew** the breadth and depth of FSBO staff and partners surrounding all aspects of financial products and services would create a new utopia of leverage and profits. Before the recession, **we knew** investments would appreciate with little downside risks. Before the recession, **we knew** the safeness and soundness of the financial systems—global and domestic—were unshakable and would never come into question. Before the recession, **we knew** many things for sure—but that is what got our stoic FSBO's into trouble, and in the process taking the Davos celebrities off a cliff of their own making.

It was a revisiting of this popular Mark Twain (i.e., Samuel Langhorne Clemens) quote that ushered in the spark for this book, and became its theme for looking at FSBO's and their operations in a world that was no longer of their design. For you see, their design (and operating beliefs) changed little from the mid-1960's even with progressive adjustments in technology, processes and regulations. What was substantially missed was the consumer changes and attitudes toward FSBO's especially as empowerment in purchasing, social interactions and brand voice rose beyond the brick and mortar of the once valued branch.

When this book was just a series of collaborative emails and texts between two people, the idea that an iterative book series could be created about changing FSBO business models seemed at best, far-fetched. Many said, even a few individuals in their 80's, that non-fictional textbooks are dead—passed into history like Myspace, the flip phone, Invader Zim, and a single purpose mobile app. In short, it was something no one would ever read.

It appeared, when we started this effort with a Tweeting POTUS in the Oval Office, our success at creating a concise, easy read, deep-dive set of constructive analysis and potential high-level recommendations for FSBO's was not advisable. We noted that there were vast numbers of daily articles, short bursts, and non-vetted blogs that filled content needs at the push of a button or a voice request (and now 2-way video) via Alexa.

However, when dealing with highly complex and uncertain regulations along with an evolving customer base, which consistently ranks FSBO's at the bottom when compared to other industries, we asked the obvious question, "Why are FSBO's, nearly a decade after the Great Recession, not resonating with any of the generational demographics below the aging Baby Boomers?" Indeed, research to answer this subject yielded many more questions surrounding market disintermediation, commoditization of offerings, regulatory challenges, digitization complexity, personnel and operating structures and additional topics summarized in the following Introduction.

We also constructed the content of this book in a collaborative and iterative process—we wrote only after reviewing numerous topical questions, orchestrated and taped conversations, and finally developed visual diagrams and tables to wrap complex points into understandable concepts.

What resulted at the end of this atypical build process was a tightly coupled starting point for a book, which reflects our nearly 60 years of "operating without a net." You will find the general text conversational, as if two industry personnel were having a working lunch trying to determine what's next, why is our current state waning, where are the opportunities, and what should we be doing to prepare ourselves and the enterprises for which we work?

Additionally, this book was not designed to be another War and Peace—we worked to create less than 80,000 words. Whereas, many business books create deep dives into topics ranging from the use of cloud computing to settlements to organizational processes, we chose to construct this conversational artifact (which is how we wrote it as a P2P dialogue) to be broad and cohesive. The reason was simple—too often writers and researchers concentrate on details without ever framing the drivers and wide-ranging changes taking place across FSBO's.

This is not to say that we don't provide solutions, or at the very least, the next steps FSBO's must take if they hope to evolve and meet the needs of this new market. We take stock of the situation along many operational lines in the first half of the book, and then look for solutions to the identified problems in the second.

And for those readers who crave more, and to demonstrate that what we are saying has efficacy for the emerging FSBO's beyond the traditional B&M (brick and mortar), we have made use of "call-out" boxes throughout the chapters to reinforce our points, offer a different perspective, or to provide "A Deeper Dive" into a supporting theme. These callouts can be skipped for a less involved and faster reading of the text, but we feel they offer the best opportunity to touch on the more complex and necessary challenges that organizations will be required to undertake.

It's worth noting, that while many business, economic and strategy books dive right into the need for technology to transform or morph a FSBO, we believe that these taxonomies of discrete solutions are just a means to an end. This book is not singularly about technology or the next "killer app" that will make your institution "the best". There are enough of those to start a library like self-help themes or diet books. This book is not about the latest fads—it is a holistic and lasting series of principles and frameworks that cut across today's innovations allowing for continual adaptation rather than a flash-in-the-pan or advocating a technology trap.

There are many a failed institution or initiative that have superior technological innovations and market facing benefits—but they failed for lack of an understanding how they served the business, and not how the business served them. It's not to say we won't discuss these topics in callout boxes or in the main text itself, but these popular themes—cloud delivery, machine learning, distributed ledger[i], cryptocurrencies, and even predatory FSBO's (i.e., FinTech)—are not the answer, they are part of the solution, not an end state to be reached and a box checked off.

Lastly, at the end of each of the ten chapters, you will find suggested priorities or challenges that most FSBO's should consider—think of it as important take-aways. Each of the illustrative priorities, ranked from one to three, are addressed using the "four-V's" mentioned in the introduction and referenced throughout the chapters. A more complete discussion of the four-V's can be found in **Appendix B**. Like the call-out boxes, these lists are provided for those seeking a path-forward and potential scope boundary for various programs-of-work.

By providing a stake-in-the-ground with this book,

we set a holistic framework through which FSBO's can respond to the marketplace revolution brought forth by regulations, FinTech, private secondary markets, and younger consumer segments more focused on social responsibility than profits or brands. After having spent our entire careers in FSBO technology, c-level management, and back-office environments, we recognize the need moving forward beyond the technology utopias envisioned by architects and solution providers into a world that will be highly transitory especially when it comes to consumer technology.

To promote a little bit of levity, each section will be preceded with a cartoon about the contents to follow. After all, if we can't jest about our challenges we just might succumb to the massive amount of movement taking place through all channels of provisioning and delivery. As this initial scene portrays, it will take considerable deliberations and hard work to achieve our strategy shifts as, we are sorry to say, we don't live in a magical world.

And stating the obvious, but to be clear to you as a reader seeking directions, not all the answers are in this discussion—no matter how much we edit. We have provided the foundations and frameworks to move forward, while avoiding common traps, but the totality of and complexity of the FSBO world is vast, and no single prescription will cure all ills.

And finally, it has been a decade since your authors last collaborated. We trust it won't be another before we create the next installment.

Mark Dangelo *Rick Grant*

"I think it's important to note that we really did try hard."

Image ©Shutterstock, Inc.

Introduction

"If the American people ever allow private banks to control the issue of their currency, first by inflation, then by deflation, the banks ... will deprive the people of all property until their children wake-up homeless on the continent their fathers conquered ... The issuing power should be taken from the banks and restored to the people, to whom it properly belongs." ... "Paper is poverty. It is the ghost of money and not money itself."
– **Thomas Jefferson**

We've both occasionally wondered, as we approach the culmination of our business careers, how many professional books we have read? What made the two or three we can probably remember—if we can even remember their titles—memorable as compared to the others? Were they thought provoking, controversial, focused on going to war, required for a project, or concentrated on redesigning parts of the enterprise? Did they speak about technology, personnel, regulations or even the availability of data? Were the ones we remember simply the last books in the queue? In the end, did all the reading, work and results really boil down to, "We tried?"

Still, a more fundamental question might be, "Why does the world need yet another transformation book?" As we know, regulatory compliance has changed the face of financial services and banking organizations (FSBO's) since the Great Recession[ii] of 2008. Additionally, since the recession there has been a four-fold increase in FSBO customers within the top 5 institutions as ranked by asset size—these institutions now control nearly 50% of all accounts.

As FSBO's have become more, not less, systemically important to the domestic economic expansion, they also face fewer competitors, are viewed with increasing societal distain, continue to commit transgressions for profitability and are perceived as impediments of innovation due to their cultures, which seem dedicated to greed.

These probing questions and the perceptions consumers have developed of pervasive bad actors within FSBO's have effectively reduced the barriers to entry for startups and crossover firms (e.g., FinTech), allowing them to create beachheads in an industry long dominated by conservative leaders. Given the challenges the industry faces today, along with what we see transpiring within the secondary markets for mortgage securities, we find traditional FSBO's falling into one of two distinct operating models—they choose to become a utility or they learn to become customer facing.

The utility model brings with it extensive regulation, and requires a high-level of capital investment. The result mirrors our national utility grid, with the utility generating the power and the lines distributing it to homes. This model, and its new iteration labeled BaaS (Banking-as-a-Service) or BaaP (Banking-as-a-Platform), has been enabled by not only mobile technologies and partnerships with consumer app developers, but also through the acceptance of standards for a fast-growing ecosystem of API's (estimated at over 15,000 and counting). The utility model provides the scale needed to sustain startups, while satisfying a consumption versus build deployment demand.

The idea of building a "killer app" provided the genesis for FSBO's to partner with consumer facing financial solutions and social media providers. Post-Great Recession, FSBO's were directed by investors, boards, and regulators to concentrate on compliance, leverage ratios, and living wills to meet public and private demands (i.e., penance) for safeness and soundness across the financial services supply chain. The result was a bifurcation and a relinquishment of market segments to those better able to provide consumer services—as compared to high-capital build demands of middle and back-office solutions.

So, these are a few of the drivers that made the case for creating a different form of book for FSBO's. The markets are splitting and it is very likely that in just three to five years the fork in the road that created the events of today will each branch into additional forks. These new splits will diverge to a niche play or merge back into a common model, depending upon the speed with which FSBO's accept and adapt to the revolutionary environment in which they are now competing.

Combine these events with highly divisional and rapidly changing political messages and FSBO find themselves in an environment in which their power to hold captive their consumers has dissipated. Consumers now have the power to quickly change institutions with little pain or financial loss. The power is now in the hands of the account holder—not the provider.

Banking is the cornerstone of capitalism—without access to credit our societies would stifle and fail. Even communism came to understand that leverage and payments can make or break a socio-economic structure. We believe that the day of paper currencies is over, receding into the past. We believe that T+0[iii] is not a goal; it could have been achieved, but only at the risk of significantly decreased profits for major institutions. We firmly believe that traditional credit cards, most brick and mortar branches and traditional loan processes will vanish within the next decade, to be replaced by the digital equivalents used by a mobile society. We even firmly believe the idea of offshore outsourcing's relevancy (e.g., ITO) has passed for a multitude of business, security and privacy reasons.

Yet, in the battle to be "like(d)", FSBO's are frequently substituting instant gratifications for longer-term demands that include competition, extraordinary unforeseen risks, and the challenges of a globally interconnected financial system (and there are many, stacked layer upon layer) which demand specialized expertise and sophisticated technologies. Like an ambassador seeking to influence a foreign country, we can only bring our concerns and recommendations to the varied presidents of the FSBO's in the hope that being popular is not more important than being ethically successful—but there will be negative episodes in the absence of action.

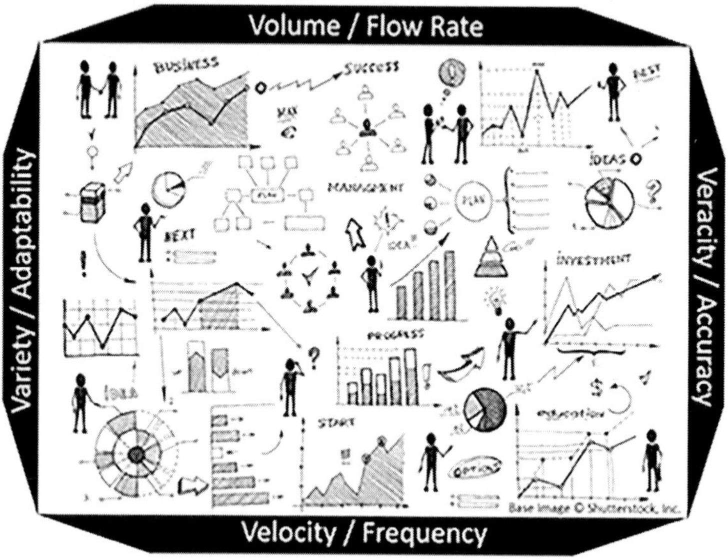

Therefore, to provide a viable plan of attack to adjust or even radically transform an enterprise, we have divided this book into two main sections all the while encasing the problems and solutions around the "4-V's" (see diagram below). The first half of the book (i.e., chapters 1 to 5) is concentrated on why permanent and radical changes are happening and the second half (i.e., chapters 6 to 10) on what must be accomplished and internalized to ensure evolution as information becomes known and results achieved. The appendices are additional in-depth discussions to aid delivery concentrating around the use of big data and complex, layered digitization—all required competencies for any FSBO competing with non-traditional upstarts.

In an era of populism, it seems like every bit of assistance or advice is focused on slogans and divisions. What FSBO's need are capabilities that allow them to execute against an innovatively relevant business model that is both adoptable and adaptable. What is provided in the following pages starts with a purpose, a burning platform if you will, and culminates with a framework that can be molded to meet the specific challenges facing your FSBO.

This will be the fourth major cycle we've worked through as an industry. In this cycle, technology investment becomes so commonplace that all new FSBO solutions touted or brought to market can only be labeled "different."

The solutions we typically see today are advanced regarding consumer technology and processes stacks. They fall short of creating digital journey maps which bind brand, solutions, talent, investment, technology and results into an innovatively relevant cohesive operating model. Being a little cheeky and contrary to some "widely held beliefs", an excessive tweeting CEO is no better than one who is more reserved—tweeting is not by itself a FSBO transformation, policy program or governance strategy.

Edward Deming noted decades ago that the use of slogans only creates internal divisions and an operating environment of blame (see Point 10 of his "14 Points for Western Management"). Within FSBO's spurred by the accelerated change brought forth by technology and regulation, programs and projects long on rhetoric generally yield little in terms of sustainable results.

Hence, this initial book was envisioned, designed, and created to look beyond the populism dominating the business publishing industry (i.e., like a diet constantly being proposed to lose weight if you give us additional money).

We provide a framework—not a rigid prescription. It's a fresh approach to the countless also-rans pushed by some consultant's solution of the year (e.g., how many of you ever were told by advisors to read "*The Art of War*[iv]" only to have them a year later push another slogan based approach?).

We hope to offer you much more than that with this work—and indeed like an open source solution, we invite you to improve upon it and make it yours.

However, if you must have a slogan, remember, for FSBO's to prosper, they need to move **Beyond the Technology Traps.**

NOTES

"And that is what happens when we resist change."

1

A Problem with Profitability

"The banking business is no favorite of ours. When assets are twenty times equity—a common ratio in this industry—mistakes that involve only a small portion of assets can destroy a major portion of equity. And mistakes have been the rule rather than the exception at many major banks." – **Warren Buffett**, Chairman and CEO of Berkshire Hathaway[v].

In 2016, U.S. voters elected Donald Trump, a businessman with no political experience, to the highest office in the land. In the process, FSBO's (Financial Services and Banking Organization's), which had witnessed nothing but legislative and regulatory changes since the Great Recession a decade earlier, found themselves teetering between more of the same and a new administration that promised to roll back regulatory control of American business.

While we are not convinced yet that any changes will be forthcoming, aside from those which are driven by trends that have been gaining strength for many years. In our view, encompassing the larger picture of FSBO as a collection of varied industries spread across a financial services supply chain, regulatory oversight is but one strand in a web of oversight that threatens the financial services industry when it comes to competing across globally connected markets and a diversified customer base.

Regardless of the changes this or future

administrations envision in the coming years, the past 30 years of global market interconnectedness and layering, which has created a business environment that transcends geographic boundaries and localized consumer behaviors, will not likely be undone. Moreover, taking a view that less regulation is better will not erase the negative public perception the industry earned through the historic downturn and that bad industry actors continue to deserve.

Worth noting in the weeks after the 2016 national elections, the U.S. President was already backing away from some of his most aggressive campaign promises. His first commercial action, a negotiation with Carrier—a firm that had planned to send thousands of jobs to Mexico—succeeded only at the expense of hundreds of millions in tax concessions. There is reason to believe that the changes voters had hoped to see will not materialize, at least not in the manner they had hoped as local and state pressures out-trump federal wishes. As this book was going to press, the new administration had been unsuccessful in creating any significant change in its first seven months in office.

But this book won't be about politics, though that is certainly a force we will consider as an impact to domestic FSBO firms operating in the U.S. as well as globally. Instead, we plan to discuss many threats and opportunities that face executives working to lead mortgage lending[vi], and in general, FSBO companies over the next few years. The new administration and its decisions will have an impact, to be sure, but so will demographics, technology, consumer sentiment toward the industry and competition to touch on a few of the stratospheric topics. We'll discuss these and more discrete ones, such as financial supply chains, marketing, disintermediation, commoditization and "what's next."

Where Did We Start?

We (your authors) each entered the FSBO industries over three decades ago. When we began writing about FSBO's, and our experiences within their highly siloed operations (i.e., before Gramm Leach Bliley[vii]), the industry had been operating on culturally infused business as usual (BAU) for decades. While there were new technologies introduced to the market in each year prior to the Great Recession (2007-2009), they were largely tools to automate existing manual processes. There was little if any significant supply-chain reengineering, and delivery systems were always viewed as a castle, requiring tens of thousands of pages of documentation to understand and taking years to define and implement.

Doing the same old things in an automated fashion, fueled by the government's desire to increase homeownership, led to a proliferation in loan originations that grew in the early years of this century to historic proportions (e.g., home ownership > 70%, over $12 trillion in mortgage debt, >60% of all loans with some form of speculation, or > 100% of market valuation)[viii].

When those bubbles burst in the latter part of the millennium's first decade, the U.S. and many parts of the connected world economies fell into a massive recession that most Americans still blame on the greed FSBO's built into their cultures. When the U.S. government bailed out the nation's largest banks (and even some of the global institutions that had sizable domestic operations), leaving the middle class to count the losses in its retirement accounts, the reputation of FSBO's were severely damaged, perhaps beyond repair.

With the passage of the Dodd-Frank Wall Street Reform and Consumer Protection Act, an industry that had operated each year as it had the year before, entered a new age under the oversight of a new federal regulator. The Consumer Financial Protection Bureau (CFPB) started out with a list of hundreds of rules to write under the authority of the new legislation. The creation and flushing out of discrete rules has been ongoing since the Bureau's formation in 2010 (and it is worth noting that the CFPB, for the most part, writes and enforces their own rules as compared to other oversight agencies). Almost overnight, regulatory compliance became THE most important issue for bankers, more so than balancing other risks or finding the key to making money from an increasingly discrete and changing consumer base.

Even though a new presidential administration has vowed to force regulators to delete two regulations for every new one they write—a mandate that seems impossible to enforce and arbitrary in its impact on the industries under regulation—FSBO's cannot expect to return to the good old days of finance with their "celebrity appearances" at Davos.

Banks operating in the current and future environments must be compliant, but also so much more. Ultimately, everything covered in this book will deal with compliance, marketing, technology, and competitive strategy as common threads. None of these elements can be viewed independently of the others. It will be in their effective combination, layering, and a practice of orchestration that successful bankers will not only survive but thrive.

The call-out-box below contains a conceptual diagram that brings together the discrete items necessary to craft a go-forward roadmap, highlighting the competencies that must be either created or gained to ensure success. By mapping out not only the rationale that FSBO's need for deployment, there is an explicit recognition on the integration mandates needed for leverage and reuse of people, processes, and technologies.

Moreover, FSBO's must have the talent. This will be a significant challenge these institutions must face. First, it won't be easy to attract the talent due to the high level of consumer animosity. Second, FSBO's don't understand demographics, as evidenced by their apparent need to control all aspects of the channels and apps they use to reach customers. Executives talk about slices of markets and the creation of bespoke offers, and make that approach a marketing mandate.

The reality is it's not any longer about just creating something for a slice of the market (e.g., affluent customer, jumbo mortgages), it's about creating something that is truly unique and beneficial for each individual customer—and how the institution accomplishes that. To be exact this is the struggle FSBO's face because it requires a culture shift to a more interconnected industry where FSBO's learn to be "good" partners for those outside or even inside their own industry segment.

FinTech firms have become popular with consumers and the media because they understand this need to "slice and dice" those consumer segments down to something that approximates the individual, using technology to create solutions that appear to be customized to the customer. This is contrasted to the traditional FSBO approach of "Shun closer contact with a consumer that dislikes me and just do something similar to what my competitor does." A copycat strategy will never break an entity out of the pack—at best, it allows the FSBO to continue to operate and blend into the regulatory fabric.

A Deeper Decomposition: Framing the Organization for Life Beyond Bricks

Whereas, large financial institutions and lenders have traditionally capitalized on first mover advantages and investment intensive mega-trends (e.g., back-office automation, end-to-end integrations, and football field-sized server farms), mid-sized institutions were left to pick and choose among emerging opportunities due to the high costs involved in the serial and laborious methods of implementation.

As mobility took hold and data availability has grown, the need to leverage asset investment spawned faster provisioning solutions, cross enterprise data linkages, and methods for finding the nuggets of data necessary for an accurate representation (and prediction) of how consumers and homeowners will behave. With available cash reserves on the decline as earning shrink and growth of all types slows[i], financial enterprises are seeking a greater impact from their yearly budgets.

New, compartmentalized, building block solution sets have changed the "delivery math" when it comes to market influence and domination (e.g., blockchain, distributed ledger, machine learning). By making good choices to satisfy prioritized requirements, mid-tier financial institutions can now quickly assemble and integrate complex, cross-domain apps and services at much lower costs, and without the traditional years of lead time once demanded.

Success depends upon the right set of choices among a growing set of possible partial solutions. An abstract illustration of the key components and critical benefits possible by utilizing a substitution strategy (i.e., assembly of solutions, both loosely and tightly coupled, regardless of department or internal standard—e.g., build, buy, preferred vendor, et al), is shown in the figure below.

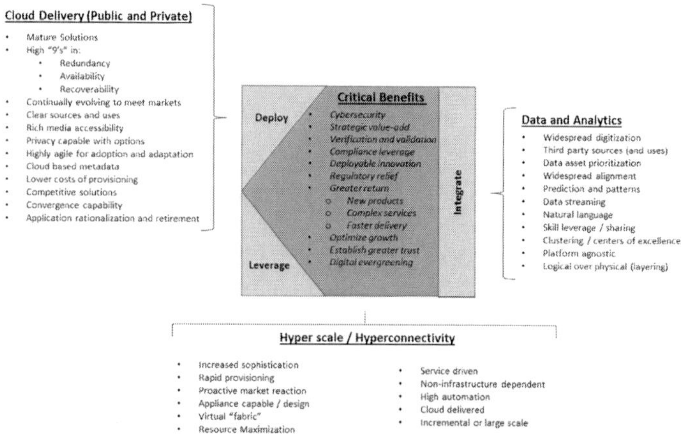

How can substitution of deployment, integration and investment leverage play a role in the rapid adoption of process-driven technology to make an FSBO more competitive? Sometimes referred to as substitution dynamics, the use of assembly swaps as a principle of complex problem solving has been around within scientific communities for decades.

These principles, when applied to technological and financial behaviour, allow institutions to realize vast competitive advantages, as well as allow them to respond to cybersecurity risk, regulatory and compliance risk, changing market demographics, operational efficiencies, and even how to profit from the endless digitization of everything.

[i] "Cash Piles at American Companies Are Shrinking", Bloomberg, OJ Renick, September 2016. "S&P 500 companies have posted negative growth for six straight quarters, a stretch that been exceeded only once since 1936..."

Why Consumers Continue to Think all FSBO's are the Same

This brings us to our first examination of a key topic within this book—commoditization. Consumers rightfully believe that "All banks are the same," and exclaim, "Oh, it's just a depository institution," or "It's just a place to get a CD or checking account." As consumers continue to view the banking industry as a place to acquire financial products that have all become commodities, traditional institutions will continue to be disintermediated from borrowers by any competitor that can offer a product which looks similar.

Ultimately, reaching the new generation of financial services customers, either as financial product prospects or as company staff members, will depend upon the FSBO's technology capability, and a clear connection between the consumer's needs and what the company must offer. Failure to take delivery of company messages through technology seriously will put FSBO's at a severe disadvantage as seventy plus percent of both prospects and existing customers believe that if the company doesn't offer an online app, it's not worth patronizing.

Over twenty percent of all current customers will never go into a branch. We've both spoken to consumers who told us they haven't been in a branch for nearly six years. Turning away from the branch model may seem extreme to institutions that were founded on the strength of the local branch during a time when over 60 percent of the population used one at least once in their life. However, today's consumers view the requirement to conduct business in a branch as arcane.

Many banks are reluctant to take focus away from branches when competitors are still building them. However, FSBO's cannot be effective followers in a market where offerings are so iterative and adaptable and coming so quickly on the heels of the last. In a market that moves that quickly, opportunities are missed, spending to meet marketing standards sets by customers increases and consumers are left disappointed. Playing to the competition instead of responding to the needs of consumers will not be an effective strategy in either the current or future markets.

To build FSBO's that can effectively compete for current and future markets[ix], a foundational stone (a very big one) must be the creation of an umbrella of digitization within the enterprise that can adapt to new requirements iteratively to impact not only company technology, but also company culture, staffing, and marketing.

Digitization must be viewed as a core competency for both horizontal and vertical integration of products, services, personnel, and partners. This can be more of an expectation challenge than it appears as digitization has been a common adhesion to bond architectures and infrastructures together for some time. It is analogous to the way FSBO's formerly approached infrastructure design. All new infrastructure was built across legacy systems by necessity, as solutions were provisioned along artificially created departments or divisional boundaries in siloed institutions.

Effective digitization across the enterprise will render every other initiative more successful and less expensive to implement. One example is the application of big data to gain understanding and predict outcomes (for an in-depth perspective, see **Appendix B**). Other examples include initiatives focused on robotic mortgage production, artificial intelligence (AI)[x], virtual reality (VR), mobility, and even the less sexy, but equally important concept of application retirement. Digitization is the glue between the systems, but it is also the nervous system that makes sharing information easier than ever before.

Ultimately, digitization offers the advantage of decoupling the paper from the product in the customer's experience. Consumers do not appreciate having documents dictate how they consume financial products and services, just as they refuse to be directed to bank branches they view as completely outdated (not including kiosks and ATM's) to conduct that business. The concept of receiving a PDF document, printing it out, signing it with "wet ink", scanning it back in, and sending it back to the institution is viewed as one short step away from carving their initials on a stone tablet.

Today's consumers demand digital options. Unfortunately, too many consumers seeking an FSBO that proactively addresses their needs and behaviors often conclude that most, if not all bankers, are suffering from amour de soi[xi].

A Deeper Realization: The Trends are Gaining Momentum

It seems that across every headline, something negative is being communicated—presidential candidates (i.e., blue, red or green), state-of-the-financial services industries (e.g., systemic risks[i]), lobbying efforts (i.e., CCMR[ii]), government affairs (e.g., cross-border, sovereign state), and of course, regulatory oversight (e.g., Dodd-Frank repeal). Within the financial markets, a mediocre post-recession normalization has resulted in atypical lending practices, greater regulatory oversight, tight housing inventory and bankers posting profits using unnecessary fee-based income.

It gives the impression from the newsfeed communications that all is wrong with the world, especially within financial services, and as commerce-interlinked societies we are all on the brink of failure—too much debt, limited real growth, stealth economic indicators and very disgruntled constituents. It appears we are further losing multi-generational standards of living, giving our children unserviceable debt and an inability to globally compete. These doubts make it hard to get up some mornings and have faith that the cross-connected markets (e.g., bond, equity, trade) along with risk-hedged currencies will not severely fall or implode, eliminating the $90 trillion of stored wealth within the roughly 180 million American households[xii].

Whereas we cannot singularly solve society's ills, we can recognize that within our collection of industries there are numerous, but less publicized, changes occurring which will have a profound impact on profitability and market perceptions. Specifically, if we examine how technology (i.e., software hardware, services, AI, people) is provisioned, the creation and deployment of compartmentalized solution building blocks for greater value, and even the retirement of legacy systems to streamline operations and increase margins, we find that underneath there are cohesive advancements.

Within market repositioning and technological innovations, it is very likely that mid-market and community bankers may outperform their much larger counterparts over the next five years, if they embrace the principles of substitution dynamics being delivered as a result of three intersecting trends—widespread public cloud adoption for rapid adaptability, data availability[iii] using secure analytical cloud tools, and hyper-connectivity to interconnect and leverage investments.

It is at the center of this broad intersection that we find the greatest opportunities for achieving operating efficiency and improved decision making against larger capitalized competitors. Moreover, these three trends also are complemented by greater machine automation of complex, cross-platform tasks (and now supported by streaming connectivity, which further feeds cloud growth), hyper scale transformations and endless streams of data, managed[iv] by emerging analytics.

---

[i] See "**Maybe We Clamped Down Too Hard on Finance**", Barry Ritholtz, August 26, 2016, Bloomberg. "The Regulatory pendulum after the credit crisis has swung too far, in part as a result of a broad lack of

trust in Wall Street among both the public and policy makers. This is stifling the creativity and initiative essential for the vibrant financial industry needed in a modern economy."

[ii] Committee on Capital Markets Regulation. Recently they published a position paper on the potential of legal action against the U.S. Federal Reserve. The first of their papers released is "**The Administrative Procedure Act and Federal Reserve Stress Tests, Enhancing Transparency**", September 2016.

[iii] Big Data (and mobility) are estimated to hold $10 to $20 trillion of benefit by 2020, "**Risk and responsibility in a hyperconnected world: Implications for enterprises**", McKinsey and the World Economic Forum, Chinn, Kaplan and Weinberg, January 2014.

[iv] Oracle is well-capitalized and market facing on its ability to offer database expertise and recovery appliances. See one of their Showcase Solutions in "**Oracle Leverages its Database Expertise to Deliver Purpose-built Recovery Appliance**", by Buffington, Peters, and Keane. October 2015.

Digitization is Assumed—Moving Forward it is a Layering of Complex Models

Digitization, as a competency, has moved well beyond the primitive states of electronic records and transactions. Fundamental digitization capabilities and implementation successes are now required core competencies for success, as FSBO's will find themselves unable to compete in a market that is seeing requirements move from simple to complex layering and dynamic stacking of views. Paper served up by branches will be abandoned by a consumer obsessed with digital content delivery and multi-threaded personal and social interactions.

If we think of digitization as the platform, a baseline of constantly changing electronic building materials, we witness the creation of a new kind of financial institution. One capable of not only understanding the information it has but also able to assemble future solutions rapidly, with superior accuracy, while efficiently disassembling existing solutions for either retirement or modification.

Among the most critical challenges FSBO's face is the recruitment of qualified staff members who can deal with advanced digitization demands and who have the market intelligence to know the requirements, both in terms of internal assets and investments, and in terms of external partnerships, for delivering seamless results in an on-demand world. This will be critical because without this competency inside the FSBO enterprise, connection to the financial supply chain will become much more difficult, as it is now built upon compartmentalized digitization capabilities and building blocks.

Assembling the right platform or discrete solution will continue to be a challenge for the FSBO. The company cannot hope to create every piece of it, which will require a layered or assembled approach. This is no different than the challenge faced by any manufacturer who relies on suppliers to build parts for assembly into end products—automobiles, electronics, computers, home systems, et al. What makes the final product different or value-added is how they are assembled. For FSBO's this means the software that tightly couples the various building blocks together and then the service provided by the institution after delivery to ensure loyalty and consumer satisfaction.

As these new financial services offerings are created, companies will begin to focus on ownership of the resulting combination of parts, many of which will fall under patent protection. Since 2010, we have seen more FSBO's embark on patent and trademark buying sprees, as a tactic to protect intellectual capital as well as provide lockup or barriers to entry to mitigate competition and ensure substantial revenue streams.

In fact, we have seen the "Big-5[xiii]" FSBO's taking the outputs from the heavy investments made in their innovation labs to protect their four-fold market share increase of the last six years. These institutions are patenting solutions that span blockchains to data interchanges to AI to VR. To stay ahead of technological changes, these oversized TBTF institutions turned internal innovators loose to create anything and everything that could cement consumer loyalty, while promoting to the regulators their conformity to rules.

Whether it works out for the big banks or not, there now exist defensive market barriers that will make competitors work harder or force them to pay larger players for technology.

There are many "pieces and parts" to consider in the process of evolving traditional FSBO operations into something a consumer might believe adds value. The process of digitization will be the foundation linking social platforms to retail platforms to financial platforms.

A Deeper Conversion: The Intersection of Agnostic Segments

Mid-tier and community banks are actively seeking a reprieve from regulations designed for much larger institutions (i.e., the TBTF and systemically important), and the acceptance and incorporation of substitution assembly provides alternatives that weren't available five years ago. Rather than being bogged down with methodological biases, substitution assembly can provide repeatable rapid results, while meeting current, and we do mean transitory, consumer demands for efficacy, trustworthy and of course in the current Millennial slang, "sick."

Additionally, with the rapid changes taking place across consumer preferences, enterprises are finding their best options are to deploy layered solutions across the cloud not only for provisioning, but to adopt to a look and feel spanning an increasingly mobile consumer and emerging instant payments[i]. Increasingly, sophisticated cloud offerings coupled with provisioning shifts, are accelerating the breadth and depth of services—a fundamental transformation is occurring from building infrastructure to consuming IT as a service[ii]. Two rudimentary projections on the growth of cloud solution providers shows that server and storage consumption of all hardware at 65% and 60% respectively, with the former having a CAGR of 23% and the latter having a CAGR of 47%[iii].

Factoring in all the capabilities offered by maturing cloud

offerings, the progression is being propelled by cutting-edge hyper-connectivity solutions allowing for the creation of a virtual woven "fabric" of capabilities (e.g., hardware, software, services)—maximizing resources, delivering cross application automation and independent non-legacy infrastructures[iv]. With cloud economic models being refined using hyper-connectivity, astute institutions will find efficiency and capability across a spectrum of sophistication, machine learning, and appliance defined. Yet, the larger opportunity for many institutions will be the application of risk and its treatment using third-party offerings designed to isolate potential contaminations and exposures.

So how will data availability continue to evolve and transform financial services offerings? There are many new and evolving answers—customer efficacy, market share, margin improvement, bespoke delivery, risks, profitability, segmentations, adoptions, compliance, and hundreds of others. It has been written before that data (and analytics of the data along with advanced machine learning) are indeed the new financial currency.

Yet, like hard currencies and cybercurrencies, the exchange rate on segments—internal, third party, cloud metadata, predictive, streaming—all have value at the same time, with inherent risk of losses, demanding a hedging strategy. The inability to find the precise value for a given market leads to apprehension and horror at the very idea of trusting the results. However, when properly blended with the two other trends of hyper-connectivity and cloud services, confidence intervals can be expanded and insight gained beyond merely seeking the white spaces across dark data[v].

---

[i] See Zelle, "U.S, bank group to launch Zelle instant payments brand in 2017", Reuters, August 2016.
[ii] See "IT as a service: From build to consume" (ITaaS), McKinsey & Company, September 2016.
[iii] Ibid, source IDC and Mckinsey
[iv] "Ten things hyper convergence can do for you", HyperConverged.org,
[v] For a tangential and more robust understanding of how this might be possible, see "Risk management: Risks are lurking everywhere", iSecureMag.com, Zoran Lalic.

It's Chess versus Checkers

What FSBO's need to accept today is that they are competing in a marketplace made up of an ever-changing mosaic of options, solutions, customer demands, and partner capabilities—all tempered by regulators and fickle politicians. This is how we connect to the world, by moving within defined limits and anticipating the shifts on this game board in terms of delivery ebbs and flows.

Yet, if we have learned anything from the rise of FinTech firms over the last decade and the massive shifts in the way business process outsourcers are delivering on an emerging process, it is that not everyone is competing under the same set of rules or with the same methods. It is as if some are playing checkers and others are playing chess. Consumers expect FSBO's to be more flexible in how they deliver, more strategic in how they deploy technologies to meet emerging customer needs and more capable of anticipating those needs. That's chess—thinking several moves ahead as the match continues. Too many FSBO's are still playing checkers, making the same moves and expecting consumers and competitors to play by the same rules. It should come as no surprise that they are losing the game.

We expect to see FSBO's make some serious mistakes as they attempt to leverage the efforts returned by their innovation labs. These institutions have a history of myopia and hubris that makes them believe that they alone can chart a path through the unknowns and past the sinkholes of risk. We've seen this before.

Back in the late 1990's, FSBO's (then S&L's, community banks and the top 15 institutions, less their investment banking offerings), adopted a similar focus on front, middle and back-office efforts as they transformed from IBM 3270 character-based mainframe dialogues to client-server approaches[xiv], non-serial delivery approaches, expansions beyond refrigerator-sized disk drives all to internalize rapid technological expansion.

This "knee of the curve" event is happening again in what some have called the fourth computer revolution. The new market changes now require a quantum shift in the company's content and capabilities[xv] in terms of education, training, staffing, partners and even co-operative arrangements. As in prior, non-stepwise shifts (i.e., non-incremental), it is the people inside the FSBO who are going to make a ton of difference, maybe all the difference for some of the institutions.

So, how do we operate moving forward in the face of seemingly pervasive need for FSBO solutions that are customer driven and defined, but also highly transitory? It's like a triumvirate or a three-pronged stool. Even with the correct staffing, without the right organizational structure, management will not benefit from these people as their voices will be drowned out by the masses and discouraged by a reward structure that doesn't accept change.

These people, in turn, will become disillusioned and join the consumer chorus of FSBO animosity. For FSBO leadership, the question becomes how to pay these people and attract them when it seems that those who are qualified won't work for the industry or won't work for what the industry wants to offer. This HR landscape is a source of confusion for those with traditional hiring mindsets, when they encounter prospects who do not want to work for a FSBO, but rather are seeking a position in a FinTech startup in San Francisco, New York, Austin, or even Chicago. Yet, data from the last few years suggests that even this axiom of prospect competition has begun to change as new graduates swing back to stability with social responsibility in larger firms, such as General Electric or Tesla.

We see less of the lottery mentality today, not only from existing company personnel, but also those with highly coveted technology skills coming out of college. Unlike those of the last two decades, today's prospective employees seem to realize that it takes a great many rolls of the dice to land on a startup that is sustainable or even salable. A shift in the political landscape or one bad management decision and they fear they will be unable to contribute to the organization, let alone have a future. For many individuals who could become highly valuable team players, the idea of choosing the right organization becomes more and more critical, especially after seeing their parents saddled with huge debt loads[xvi], underwater on their mortgages, and handed broken families all, from their perspective, "at the hands of FSBO's."

These prospects also have one other change for FSBO's to consider. It is no longer about solving just the "greenfield" potential, but solving the "right field." It is not a stretch for FSBO's to look to other industries to see how they are dealing with quantum change. An example can be found in the various, highly complex automotive industries as they deal with the phasing out of gasoline and diesel engine sales, expected within a short 15 to 20 years. At the same time, firms in this industry are spurred on by a changing focus for both governments and consumers, which is putting pressure on them to create the "next big wave" that will impact not only the environment, but tens of millions of workers across the globe.

Even once heralded startups such as Google have become so big, so complex, that they don't have the money to throw around at purely random ideas on a whim anymore. There are still exceptions, such as Amazon and Alibaba, that invest heavily in many markets to the displeasure of investors, but they are the exception, not the rule.

Still, many FSBO's cling to the mistaken assumption that they still know their market better than anyone else. We consistently see shock and surprise in the eyes of banking executives when they learn they are not actually talking to their real customers, but rather a surrogate often controlled by a retail channel, a social platform, or a mobile solution of which they are not in total control (as they have historically been accustomed to doing). FSBO's also continue to make the same mistake of assuming regulations will somehow protect them from complete disintermediation.

As we will see in the coming chapters, going to market with commoditized offerings, artificial shields made of dogma and outdated, non-electronic delivery methods, will yield little to improve margins and investments.

As in any other business, success will depend upon responding appropriately to opportunity, while mitigating the risks that could negatively impact profitability. In the next few chapters, we'll go into detail on some of the most critical risks facing FSBO's today.

Illustrative Challenges / Opportunities

		Controls and Implications (The 4-V's)			
Cate-gory	**Illustrative Challenges / Opportunities**	**Volume/ Rate**	**Variety /Adapt**	**Velocity/ Freq.**	**Veracity/ Accuracy**
1	Determine the proper mix of offerings on a customer-by-customer basis for creating / delivering trust / value.	↑	↑	↑	↑
1	Optimize the prospects of increased regulatory burdens at the same time as dealing with the uncertainty of regulatory repeal spread over years.	↑	↑	↑	↑

		Controls and Implications (The 4-V's)			
Cate-gory	**Illustrative Challenges / Opportunities**	**Volume/ Rate**	**Variety /Adapt**	**Velocity/ Freq.**	**Veracity/ Accuracy**
1	Commoditization of offerings (and its perceived weakness) brought forth from FinTech vendors, outsourcers, and appliance models linked to banking as a utility.	↑	↑	↑	↑
1	Understand and deliver strategies for a mosaic of changes to business models and operating circumstances requiring a continual orchestration of "tiles" creating varied outcomes.	o	↑	↑	↑
2	A misunderstood operating axiom which accepts simple digitization of data as the end state missing the future demands of layers of digitization, hyper scale and compartmentalization.	↑	↑	↑	o
2	Too much confidence in existing channels and customers cemented by processes, personnel and a culture unaccustomed to rapid external competitive pressures.	o	↑	↑	↑

		Controls and Implications (The 4-V's)			
Cate-gory	**Illustrative Challenges / Opportunities**	**Volume/ Rate**	**Variety /Adapt**	**Velocity/ Freq.**	**Veracity/ Accuracy**
2	Inability to assimilate and internalize (i.e., take advantage of) the breadth and depth of data generated by transitional products and third-party linkages / alliances.	O	↑	↑	O
3	Non-traditional competition that is being met with technology in a race for equality and consumer app space.	↑	↑	O	O
3	Banking animosity and an industry distain made worse by populism and widespread options beyond traditional banking infrastructure and offerings.	O	↑	O	↑
3	Expanded offerings made possible with dynamic and virtual provisioning methods that change the behavior from "build" to "consumer"—reducing barriers to entry.	↑	O	O	O

↑ – Increasing non-linearly, unpredictable behaviors
O– Increasing but predictable

NOTES

**"Ok, here's the meeting agenda:
heal the wounds, listen to the partners,
fix the leaks, make the deal,
soothe the suppliers, spank the banks ..."**

Image ©Shutterstock, Inc.

2

The Risk of Market Animosity

"The bankers might not have said it in so many words, but gradually their strategy emerged: Target families who were already in a little trouble, lend them more money, get them entangled in high fees and astronomical interest rates, and then block the doors to the bankruptcy exit if they really got in over their heads." – **Elizabeth Warren**, Senator, Massachusetts.

The first risk the modern FSBO must face regards securing the customer. In the past, when banks were the only institutions that could serve the day-to-day financial needs of consumers, marketing was largely the work of taking banking customers away from another institution. Everyone knew that once a consumer entered the workforce they would apply for a checking account and become some bank's customer. Banks with strong marketing teams pursued these young consumers, and then worked to hold them and increase the financial product share of their wallet.

Things are not as simple today. Banks are no longer the only or even the preferred financial services solution for the American consumer. Today, consumers have many options for making electronic payments, saving or parking money, and securing loans. For many consumers, traditional banks are now the last resort as what little trust they harbored for these institutions was washed away during the financial crash and subsequent foreclosure crisis.

One measure of the consumer's distrust of the banking industry can be seen in the low homeownership[xvii] rate, currently near a 50-year low. Research commissioned by the National Association of Realtors (NAR) suggests that American consumers are suffering from "Post-foreclosure Stress Disorder." During the 2008 financial crisis, 8.7 million homeowners lost their jobs, and 9 million lost their homes to foreclosure, which has had long-lasting psychological changes in financial decision-making, according to NAR. It has also left a lasting impact on the way the children of these impacting homeowners view the industry that took their family homes.

These are the children of the Great Recession and we can expect the psychological impacts to parallel those we witnessed with the children of the Great Depression. Like the young of that previous era, Millennials witnessed harm visited upon their parents, which seems to have a more profound impact psychologically than harm to themselves.

It wasn't just homes that these consumers lost. Retirement accounts, vehicles, family vacations, and private school tuition were all washed away. According to Timothy Smeeding at the University of Wisconsin, Madison, in his paper "Income, Wealth and Debt and the Great Recession," the middle class lost 23% of its wealth during the crash. He cites the most recent release of the Survey of Consumer Finances (as of October 2012, which brought in 2010 data), stating that "middle-class families lost the most wealth in proportional terms, while those at the top had, by 2010, recovered their wealth lost in the recession. This pattern arises because middle class losses in the housing market have been substantial and persisting. Indeed, home values for the middle class (their biggest asset) dropped 30% nationwide from their 2006 peak, with little sign of recovery."

Since then, housing prices have begun to improve across the country, but the damage has been done. The impact has been an entire generation of financial services customers who now view the banking industry in a negative light that is impacting their willingness to enter new business relationships.

The willingness of today's consumers to shun an industry that they perceived has wronged them is not limited to financial services, of course. Consumers can turn on any company in any industry. Take the case of Under Armour founder and CEO Kevin Plank, for example. When Plank announced on CNBC that having Donald Trump as our next president would constitute an asset for the country, professional athletes who had been promoting the company and its products lashed out—and their fans quickly followed. Social media was used to arrange a nationwide boycott of the company's products. The final impact on the company is still unknown, but at this writing its stock was still falling.

If siding with the wrong politician can damage a company in the eyes of the consumer, how much more will be identified as the reason for the financial meltdown that cost consumers hundreds of millions of dollars' worth of retirement savings[xviii] at the same time Wall Street firms made record earnings and the nation's largest banks enjoyed a government bailout?

Bankers already know the answer to this question. Not that banks have ever ranked very high on the consumer's list of favored commercial enterprises, but in the wake of the financial crash of 2007, consumers continued to be frustrated by a system that seems to favor the financial institution over the consumer's interests.

In March 2017, The New York Times ran an article entitled "How 'Consumer Relief' After Mortgage Crisis Can Enrich Big Banks" in which it outlined how the government's rules regarding consumer relief settlements make financial institutions money, whether they actually provide any relief to consumers or just invest those funds in other FSBO's that might.

In an age when the government is actively trying to empower the consumer, stories like this shine a harsh light on the gap that still exists between what consumers want and what banks have thus far been willing or able to offer.

A Deeper Reality: A Seriously Pissed Off Consumer

John (not his real name), over the years, had opened three different credit card accounts with the largest U.S. Bank. Each of these cards offered limits ranging from $12,000 to $18,000—they were all high-end cards. To simplify his life, John called the customer service center to say, "Okay, I want to consolidate these because you guys have this great card where I can bring all the features I like into one card. I understand I will pay more for it, but I get all my perks rolled into one card."

The bank's response surprised John. They put him on hold for 5 minutes, and when they came back they told him, "Well, we can't access that other information. Those are different cards based on partner programs. You would have to reapply." His response was, "Well, I just want to roll them all together so I have one bill, not three, and I get all the benefits."

The bank informed John that what he was requesting was just not possible. He would have to reach out to each separate card desk to make any changes. To which John replied, "You guys own these three cards, and you're telling me you would have none of my information to create a new one? All I'm asking is for you to roll it into a card that's basically less robust, but that gives me a single card with all the benefits I want. You won't do that?!"

"No sir," the banker replied. "Our systems won't allow this."

John hung up the phone and shook his head. After a moment of thought he decided that the bank was stupid, and that he had taken enough of their crap. He closed all three credit card accounts.

A Fundamental Shift in the Banking Power Structure

The rise of the CFPB, created as part of the Dodd-Frank post-recessionary clampdown on "rogue bankers," ushered in a vocal wave of consumerism—an outlet for wrongdoings backed up by an ability to extract a "pound-of-flesh" with fines and sanctions (over $6 billion in fines levied since its launch in late 2011). For consumers, the CFPB has been their own personal pit bull for dealing with bad actors—real or perceived.

The result has been a transfer of power to the consumer—and away from the institution. Even with a shift of political wills, we have witnessed a fundamental transference that is going to last.

Not only does this shift put consumers in a position that will leave them feeling frustrated, it won't do anything better for the consumers working inside the institution. Bank employees are now subject to a library's worth of new regulation, very little of which was drafted by legislators or bureaucrats with actual industry experience. The rules were created without any recognition or understanding of the systems that are in use in the institution today, nor why they were developed. These systems are largely paper-based processes, which is not what the consumer wants to use. When consumers push back, empowered by their experience with a host of applications for their mobile smartphones, they want to be able to share information electronically with both companies and their peers at their discretion.

This puts FSBO employees in a difficult position. Not only must they learn to comply with all the rules handed down by anyone who has taken an interest in financial services or consumer protection at the federal, state, or local level, but they must disregard consumer expectations in the process (as some of this is counter to growing privacy protections). To add insult to injury, CFPB uses consumer satisfaction as a metric for measuring FSBO performance.

Without an understanding of that fundamental shift in how we behave, what we do, and how we act as employees within financial institutions, FSBO's find themselves bereft of power to change corporate culture, or even the processes, applications, and technology used within the institution. They are left to do business with mere rhetoric, and people quickly tire of it.

A Deeper Dive: An Identity Imbalance

It should come as no surprise that bankers believe their future rests on the incorporation of innovative technological advancements—it does. It should also be no surprise that bankers want to exceed customer expectations, by anticipating the "next" trend when it comes to delivering the right products and services. Consequently, banking as an industry is now chasing a broad spectrum of new solutions from machine intelligence to big data analytics to cognitive solutions to virtual reality branches to securing patents to formulating a consumer digital identity that provides global transcendence—yes, many are doing precisely this across all segments.

But in their passion to target profitable niche demographics, are bankers losing their way in capability to satisfy organizationally complex business models and operations—while escalating consumer animosity? To provide innovation relevance, are bankers becoming "too cool" especially when regulations are written only in the wake of negative impacts (e.g.., the last recession serves as a prime example). They are driven by political winds that blow purely along populist trendlines. As a result, future litigation and fines are likely to rise, driven by unintended outcomes of current digital solutions.

Bankers seem to be blind to public reaction to the idea of repealing recent regulations (which the public perceives as the solution to ensuring safeness and soundness of the financial supply chain). Similarly, they don't seem to notice the consumer's social indigestion brought forth by wage inequality, the underbanked, high-risk loans, credit availability, media coverage, and a belief that there were material criminal activities against consumers and society that went unpunished.

Hence, before billions are somehow made to vanish in M&A actions, write-offs, and set-aside provisions, bankers need to robustly check (and not just with regulators) on the business acumen and risks of "being cool," "not being hated," offering the "fastest solution" or devising orchestrated solutions (i.e., cloud, layered, compartmentalized) that while "socially accepted", are not "financially" bullet proof.

Consumers Wielding a Hammer

Banks and their employees are not culturally accustomed to consumers dictating process or even having as great an influence as they have been given in today's market thanks to the rapid advances they have seen in solutions and personal information delivery devices. Consumers have gained purchasing power through advanced social media, peer-to-peer platforms, collaboration portals, and even through far-reaching crowdsourcing capabilities that were once controlled by financial services and banking organizations.

Therefore, when consumers are negatively impacted or they perceive a wrong, they will take to their mobile devices to tell everyone in their sphere of influence about their experiences, and the organization at the core of their problem. It is this behavioral change in the consumption of financial service products and services that the CFPB (and other regulators) have tapped into to collect and analyze consumer experiences with "hard data" (in contrast to the CFPB's extrapolations of data used to determine discriminatory practices).

FSBO's now face seemingly random compliance risks because of the CFPB in its present incarnation. The CFPB has no problem finding and handing down judgments and making settlements with the institutions, in effect slapping them around when they don't see consumers getting what "the government" thinks they should. We don't have a lot of reason yet to believe that this is going to change any time soon within the new administration, especially since the CFPB remains very interested in how consumers describe their experience with financial services.

That problem becomes magnified when we realize that social media is making it easier and easier for people to share bad news, or even fake news, which can be very damaging to an institution. Bad news is magnified many times across social media and picked up because that's what people are keying off—banks are held in such disdain that they become the institutions that consumers love to hate. We are seeing this happen to recognized brands, not because they did something wrong, but because they didn't do something the consumer felt was right. Understanding how the "fake news" cycle works will give some FSBO's an advantage in the future.

Another factor magnifying market media and regulator intrusions is that these FSBO's have become so complex and layered that their governance capabilities are not robust enough to meet the requirements. A material contributing factor in why the CFPB[xix] is spurred into action is that the FSBO don't realize the negative implications of what's going on within their own ranks. You can see that with the huge judgment against Wells Fargo recently (and CFPB's largest against Ocwen at nearly $2 billion[xx]), a bank that tracked a great many metrics that really didn't have anything to do with the consumer, but rather were focused on the institution and how the institution believed it was making money. By not focusing on the consumer's needs and the public's ability to share negative news across its consumer base, the bank opened the door to market animosity (especially against the charge of fraudulent accounts).

Demographics will only make this worse. We have 78 million emerging consumers within the Millennial demographic alone[xxi], and their behavioral consumption has had two major impacts on banks. First, they were born with the Internet. They were born online, which means they were born doing things a different way from the way banks have traditionally done them. They were raised by parents that, overall, didn't put down many boundaries or provide much direction. So, when they hear a banker tell them exactly how they are going to proceed with a transaction, they push back. They don't like it.

The second impact is that their presence in the market has driven banks to once again launch innovation centers. While this is certainly not a bad thing, the problem we see with many of these "centers of excellence" is their inability to get anything approved. The internal politics that have traditionally made innovation within the bank a frustrating process arise as soon as a new idea begins to develop.

The fact that some FSBO's are making a big deal out of their search for innovative ideas, but without developing any, is increasing consumer frustration. This is only serving to increase distrust on the part of the consumer and drive them toward some other entity for conducting their financial business.

It's ironic that a large part of the reason that banks cannot adopt new ideas quickly is that they cannot seem to get rid of outdated ones. Even automation, which was intended to be an innovation that would allow banks to meet consumer demand faster and more easily, was implemented in a way that simply made outdated paper-based processes move somewhat more quickly, when there wasn't an error or compliance problem that threw the deal off the rails and out of automation. This has led to increased complexity in the products FSBO's offer, many of which came through acquisition. If banks could shed their outdated ideas, it's likely they could save 20% to 25% of their base costs, which they could then use to embrace some of the ideas and new products that the consumer really wants.

In the banking world, we've always talked about channels. Today, it's not about a singular, optimal channel, but determining the best mix from the many consumer delivery channels spread across varied software platforms, physical locations, device management and legacy infrastructures.

Whatever the channel is that's optimal to reach the desire consumer base, that's where the banks need to be. If you look at the fundamental shift in terms of how to communicate with customers and what their expectations are, FSBO's can steer a path through the growing diversity of information delivery and create products and services as the customer uses them. Banks have one foot on each side of the chasm and it creates a "frozen" attitude of how and where to serve the consumer best.

Just look at the things from a banking perspective. "Know the Customer" (KYC) is full of regulatory obligations. You've got traditional offerings (i.e., legacy products and services) covered by new regulations that many employees don't understand. You've got areas such as recovery capabilities if we've been mistreated you can launch formal complaints as an individual (as we've discussed surrounding the CFPB). Moreover, there are creditworthiness repositories (primarily controlled by the big-3 within the consumer credit world), which are evolving—but considering a recent breach at Equifax consumers are unlikely to forget their treatment[xxii]. When you look at what's important to the consumer across that chasm that stretches between what is offered and what they want, the consumer preferences concentrate around social capability, peer-to-peer interactions, speed (i.e., end-to-end transaction times that EXCEED their expectations).

Everything you hear in marketing to the end consumer is about fast, fast, fast. As an example, consumers may pay a little bit more in terms of fees, but they don't want to spend months getting a mortgage loan. They want banks to be able to access the information that's out there in the social profiles, in their credit cards, in the auto loans and even from their mobile platforms.

These learning (and behavioral changing) consumers want you, the FSBO, to look at it functionally, not just from an application and not to bury their concerns in a bunch of meaningless data gathering bots. When taken in its entirety of why consumers believe FSBO's don't listen to their concerns, it builds to another of the critical animosities financial users have towards banks. FSBO's may have data, but they don't know how to use it. In the end, it infuriates the consumer.

A Deeper Disruption: Trends and Implications

Let's take a step back and examine some relevant fields and developments which have a material impact on financial services not only today, but as trends and implications for the future.

- Globally there was $50 billion invested in FinTech in the most recent year with domestic estimates approaching $30 billion and rising at over 30% per year.
- Since the start of the millennium, the U.S. has lost nearly 6 million manufacturing jobs representing over 25% its workforce (it's peak was 1 in 4 jobs).
- Home ownership is at a near 50-year low coming in near 63% of the U.S. population with projections pointing to further declines, while at a full 2% below the 40-year moving average.
- Auto and personal loan delinquencies and defaults are climbing, as the economy flattens and in boom-to-bust areas (such as oil) the rates are becoming alarming.
- Compliance expansion, because of post-recessionary legislation, has tripled the amount of industry guidance, while the politicization and even demonization of banking enterprises has continued to rise (have you looked at Twitter lately?).
- Security breaches are rising in size and sophistication, while alternative payment and currencies expand consumer choices.
- Energy prices are finding their new equilibrium as the stock markets are topping out due to growth expectations for developed economies stalling—

including China.

- Real (negative) interest rates while propping up equities markets are contributing to a global stagnation as governments fixate on the U.S. dollar as a reserve currency resulting in unproven economic policies.
- Yearly R&D within the U.S. easily exceeds $500 billion and represents nearly 30% of the global investment consuming an average of over 5% top line revenue (and some industries exceed 25% of total revenue).
- And finally, tech giants are piloting financial services-like offerings leveraging their huge data volumes compiled on consumers, businesses, and government segments (down to the individual, with Amazon as a new-age bank?).

Put these macro events into play around discrete bespoke digital banking offerings and the result usually is what defines to the public the organizational brand or a major component of its future identity. Nonetheless, whereas offerings are often the source of buzz, failings will generate more notoriety (e.g., fraud committed within the SWIFT network after decades of proven success). No one talks about the cool of traditional checking, they talk about mobile deposits, easy moving funds and electronic payments via Starbucks-like apps—they speak often about their dislike for traditional banks and the leaders within.

The amount of innovatively relevant technology advancements is indeed creating an "open" and layered series of interconnected banking systems spread across non-traditional partners, which is untested (and unpredictable) in times of crisis. We must ask, what does this do for safeness and soundness when it comes to FSBO's and the financial supply chains? Whereas the Fed's (i.e., U.S. Federal Reserve) stress tests are getting more stringent and focused, regulators across every segment of the financial chain are suffering from uncertainty and a global lack of deep skill sets.

An Error in Strategy: Finding a New Base

Animosity in securing credit, especially for new financial services consumers, is one of significant and lasting frustration. Of those consumers who want credit today[xxiii], many won't get it because the banks use an old system that looks at past credit history to assess the risk of opening a new account.

Without a history, the consumer is generally judged a high risk and the credit request is denied. Because of this treatment, many Millennials decided to forgo the FSBO entirely, choosing a peer-to-peer type of lender or platform instead. They opted to go to a provider that would allow them to get started with credit[xxiv] and make the purchases they desired, even if that included required short-term loan. This is business the banks just gave away.

Fundamentally, new, underbanked, and emerging consumers are turning the bank's own rules and guidance against them. Moreover, it reinforces the consumers mistrust of the entire system. They know full well that there are other options, including plenty of non-FSBO's which are not governed by the same restrictive guidelines.

Consumers know that these new competitors see them differently and are willing to help them. They have told consumers as much and then given them reason to believe it. FSBO's are retreating from the field of competitive battle, yielding to others who are willing to take the calculated risk to win a new long-term consumer.

A Deeper Reality: Strategic Endeavors are Not a Dash Button

Today, strategy is like an Amazon Dash Button—granular and seemingly usable (e.g., guaranteed delivery) with the push of a button. There are many common strategies available from media and in academic works—organic, custom, novel, social, branded, mobile, cycle-driven—and those who envision implementation (using strategy) believe that once vision and goals are set, the rest becomes a top-down series of decrees. So, why not invoke traditional or popular approaches to ensure success?

Still, with all the variability, abuse and scrutiny that strategies can be subjected against, the need for successful approaches have never been greater—hence the popular resurgence. However, the same leaders and sponsors who desperately demand results against their expensively developed goals and objectives, play the "Game of Maneuvers" attempting to ensure that their empire has what it needs, while other enterprise teams are left to chance.

Asking why this game of maneuvering is being played within the enterprise is equally as important as eliminating resistance (i.e., compromise or squash). Within the multiple answers to relevant questions resides the real reasons why associates are being played one against the other. For instance:

- Is it because the goals and objectives are destined to fail and personnel want to make certain that it is not because of their efforts (i.e., a "splatter effect")?
- Are the measurements so off-target that they have little hope of realization?
- Does the sponsor, their agents or lead divisions evoke a lack of trustworthiness?
- Have there been prior missteps or retribution between the dependent parties?
- Is there retribution associated with failure and risk taking?
- Are there rewards given to those who succeed on the backs of others (i.e., work up, kick down)?
- Are individual contributors placed above those who embrace collaboration and orchestration to aid strategy implementation?
- Are the strategies unable to be internalized by existing personnel given their education and training?

- Are the strategies too esoteric, too fluffy, or too distant (stretch)?
- Will the implement of strategies shift decision-making power or budgets within the enterprise?
- Are the reigns of proven managers being threatened as implications are uncovered (i.e., principles of operation decomposed into clear rationale and implementation implications)?

The previous items are but a few of many challenges to change and understand. Yet, it takes just one of them left unaddressed to put the enterprise at risk and strategy implementation as a lever for political action (e.g., blocking, changing the discussion, stoppage, realignments, changing of responsibility, win-lose).

However, as in all fight-to-the-death tournaments, the best methods to neuter the competitors are with truly unique strategies fully developed within the enterprise (i.e., not shifted and lifted at the hands of outside regents). There are recent parallels across enterprises like Google, Telsa, and Amazon to cite a few of dozens. The latter, with their buttons, drones, virtual services, and even visions of a floating drone warehouse, has within their market segments put the kingdoms of old into a small corner (e.g., UPS, USPS, HP, IBM, Walmart, Kroger)—where they have a choice to adapt (e.g., USPS delivering Amazon packages on Sundays) or perish.

A People Challenge Results in a Reinvention Void

The internal challenges are going to be just as difficult because recruiting good people in an environment in which most consumers have been conditioned, over the last decade, to hate them is not easy. It's going to get harder as competitors and regulators confront these dogmatic cultures.

Starting before the pre-crisis which began in 2007, a growing set of evidentiary statistics of recruitment and applicant desire to work for industry groups showed that three in ten were MBA's seeking to "make it big" in an industry dominated by oversized profits and personalities. These driven, future business leaders believed that they could be part of something like an FSBO and at the same time be individuals able to earn high-paying salaries.

Today, it's worth noting that job seekers increasingly don't want to work for mainline banks or even the future FSBO's (less than 20% of all newly minted MBA's[xxv]). These candidates have the skills, the talent, the education, the drive, but they don't want to be bankers. They look at the livelihood or the expectancy of being at these institutions for a long period of time (> 3 to 5 years), and view their career opportunities as unacceptable. They are seeking more stability and less of the offensive stigma (i.e., greedy bankers).

Even since 2010 at the start of the FSBO industry recovery, banks have reduced their total staff by 6%. That is off the 2007 watermark numbers, which are showing that over the last decade, FSBO's have lost more than 25% of their talent base. In examining these trends, the conclusion is inevitable—we're seeing a huge talent skill drain out of banking and financial services.

When everything is becoming increasingly digitally based coupled with the information of big data, and information intensive applications that cross consumer use, how are FSBO's supposed to transform their operations to be bespoke, yet efficient? Institutions don't have the skills nor the working capital (due to a decade of asset recapitalization), and now face thinner margins due to lower volumes, and competition from non-traditional firms previously unheard of in their sphere of competition. FSBO's need to be fast and nimble with the resources they have—rather than attempting to remain stuck in a history in which they controlled the outcomes of financial consumers.

A Deeper Dive: Asking Why Skills Are Necessary?

With any adopted innovation(s), be it point based, solution based or (white) relabeled, the on-going and healthy rethinking of FSBO needs to continually ask and continuously challenge the "cool" solution's principles, rationale and implications from concept to development to introduction to maintenance to retirement (likely within 1 to 3 years of introduction):

- How will the introduction, or lack of, this relevant innovation impacts our operational models?
- What are the financial, reputational, market, investor, regulatory and political exposures created? How can they be modeled and integrated with vast data throughout the delivery?
- What will adaptation do to the safe and soundness of the innovations, its operating scenarios, or its coexistence with legacy infrastructures?
- Will this innovation allow for survival in an industry undergoing disruption, or will it result in operating segmentation that cannot be integrated or adapted beyond the introduction?
- With a strong correlation between "newness" and third-party partners, can utility-like operations be created to mitigate dependencies, while ensuring control along the life-cycle development, delivery and data?
- With machine intelligence growing non-linearly, what will be the likely disintermediation results along the financial services supply chain?

- Who possesses the use cases for the "cool" FinTech and with what experiential foundations are these projection and results bounded?

This is just the start of moving beyond the market or brand "coolness" into the reality that financial series delivery is not social media, a game or even a virtual reality solution—they are merely relevant innovations that must be folded into heavily-regulated and politicized industry segments.

For any banker rethinking what FSBO is—against all the changing behaviors—it is easy to latch on to the "cool" as it is easy and expected. However, something about lemmings comes to mind.

The Likely Impact on Traditional FSBO's and Their Cultures

From the FSBO standpoint, the real risk is that the traditional bank becomes unimportant to the new banking consumer as well as to the decaying base (e.g., Baby Boomers) that once built these institutions' vast empires. This is not to say that they're won't remain important—strategy, technology, delivery processes, applications, branches—for certain demographics, but the new set of institutional lenders and borrowers are saying quite clearly, "We just don't want to be burdened with your given process. We don't want the burden of all your costs. We don't want to be burdened with all your rules and regulations." The animosity comes in when the consumer is not given options that meet their lifestyle and living situations.

The risk is that by ignoring the way consumers and borrowers want to conduct their finances, they're making themselves obsolete with an attitude—real or just perceived by consumers—of not caring about anything but profits. Why is, for instance, one of the largest lenders in the mortgage world, Quicken Loans, on par with Wells Fargo and JP Morgan? They can bring branded solutions like Rocket Mortgage[xxvi] to bear. Consumers can get qualified for a home mortgage in a "matter of minutes" versus a matter of hours or days. By not adapting to behavioral and market changes, traditional FSBO's have fallen prey to the risk of being displaced, resulting in comprehensive drain on their margins and market share.

Another challenge for traditional FSBO's stems from the way they provision and use technology. This is putting them constantly behind the curve. Within the traditional institution, for those of us who started our careers with them, standard operating procedure (SOP) was to always define, define, define, create, create, create, test, test, test, prove, prove, prove.

If there was a need for new system provisioning whether or not it was a competitive response, the SOP had to run even the most minor of adjustments or additions up the hierarchy. By deploying within that kind of a mindset and organizational culture, the obvious result is you cannot be nimble and quick. You cannot adapt let alone adopt anything "innovative." You cannot take advantage of new advancements because, "they were not invented here."

For FSBO's dealing with loan margin declines, declining traditional product mix acceptance, and an increasingly mobile, irregular series of consumer behaviors[xxvii], the achievement of efficacy when determining and deploying solutions demanded by highly segmented consumers (i.e., beyond slogans and words—booked results) cannot be unprotected and slain by political selfishness and manipulation.

With nearly 150 million business and economic books penned, it is safe to say there are thousands of texts written on why some enterprise initiatives (and the enterprises themselves) fail and others succeed. Some lines of attack indeed are flat out hideous or flawed. Other approaches are too generic and others still too confining.

Additionally, as the outreach and loyalty to targeted customer bases quickly shifts commonly underpinned by robust analytics[xxviii] (spread across virtual data sources), the performance measurement of what is deemed as a success allows proactive leaders to make mid-course corrections or abandon their deployments all together—if they so choose.

If you don't like the people that you're doing banking with, guess what? Go find one of the apps that allows you to convert and get away from a standard checking account, standard savings account, which used to be the big banking draws. A lot of that is going on and changing. FSBO's are fighting competitive pressures with technology. Where FSBO's haven't shifted significantly are their skills, their culture, and their movement beyond axioms that "just aren't so."

Illustrative Challenges / Opportunities

Category	Illustrative Challenges / Opportunities	Controls and Implications (The 4-V's) Volume/ Rate	Variety /Adapt	Velocity/ Freq.	Veracity/ Accuracy
1	Adhering to principles that FSBO's must control channels, enterprises are decreasing their margins by 15% to 25% not leveraging consumer demands.	↑	↑	↑	↑
1	Up to half of existing revenue streams will be in flux over the next five years as commoditization and disintermediation overtakes dogmatic enterprises.	↑	↑	↑	↑

Cate-gory	Illustrative Challenges / Opportunities	Controls and Implications (The 4-V's) Volume/ Rate	Variety /Adapt	Velocity/ Freq.	Veracity/ Accuracy
1	Organizational structures and operating philosophies are married to traditional cultures, which are reluctant to adopt and adapt to new delivery and transformative demands.	↑	↑	↑	↑
1	Consumers are shunning a face-to-face model in favor of mobile and asymmetrical touchpoints leaving many institutions with brand and operating handicaps.	↑	↑	↑	↑
2	Capital allocations will continue to be difficult to increase and justify as markets move towards BaaP / BaaS utility models augmented by software appliances.	o	↑	↑	↑
2	FSBO structures and incentives are misaligned for rapid transformative technology environment that brokers more in society benefit, mobility and P2P interactions.	o	↑	↑	↑

		Controls and Implications (The 4-V's)			
Category	**Illustrative Challenges / Opportunities**	**Volume/ Rate**	**Variety /Adapt**	**Velocity/ Freq.**	**Veracity/ Accuracy**
2	Micro-segmentation coupled with third-party apps now demand complete compartmentalization and layering of banking offerings—with cradle to grave under 18 months.	🡩	🡩	🡩	O
3	Channel power has permanently shifted to the consumer resulting in an opti-channel demand, which traditional banks cannot deliver without retail partners.	🡩	🡩	🡩	O
3	Consumers under 38 overwhelming distrust banks viewing them as out-of-touch, elitist, innovation laggards, and unwilling to help consumers.	🡩	🡩	O	O

🡩 – Increasing non-linearly, unpredictable behaviors
O– Increasing but predictable

NOTES

"Well, now we know what not to do."

Image ©Shutterstock, Inc.

3

The Risk of Disintermediation

"You build on failure. You use it as a stepping stone. Close the door on the past. You don't try to forget the mistakes, but you don't dwell on it. You don't let it have any of your energy, or any of your time, or any of your space." – **Johnny Cash**

The reason why market disruption works is that entrenched companies working within an industry ripe for disruption never see it coming. Even when they have every reason to believe that it will occur, they don't believe it until it is too late. In complicated industries subject to strict government oversight, this problem is exacerbated because incumbents feel that the barriers to entry are simply too high for anyone without as much experience as they possess. There are too many examples of how wrong this thinking can be to list.

Before the change of the millennium, Google's Hal Varian put forth, "A simple way to forecast the future is to look at what rich people have today; middle-income people will have something equivalent in 10 years, and poor people will have it in an additional decade." For FSBO's spurred by rapid digitization and smart mobile innovations, the time for mass disruption is arriving quicker than projected.

Why? The justification is simple. What would happen if 30% of an FSBO's customer base turned over every 18 months? To satisfy rapid-cycle and amorphous financial customer needs, what products, services, channels, technologies, processes and partners will be needed in 3, 6, 12 or 24 months? How will you retire these short-term legacy offerings, while mapping seamless transition paths all along ensuring privacy and compliance? Finally, what partnerships, investments, skill sets, and due diligence are necessary to promote success?

Disintermediation is a critical transformer for the next decade. It is continually mentioned in politics, across supply chains, within trade agreements, by consumer advocates, by fulfillers of payment processing, and increasingly in regards to the adoption of virtual reality (VR) and artificial intelligence (AI). As the world's information becomes increasingly digital—records, cash, settlement, IoT—the world that once concentrated financial profiles in the back-offices of a few FSBO's are increasingly being dispersed, while becoming more diverse and far more fungible spread among established institutions and startups.

Both of us have been to numerous FSBO industry conferences over several decades and visited globally with bankers who do not take the risk of market disruption seriously—even today. Even so, we are completely convinced that a new breed of financial services provider is even now making plans that will ultimately result in the disintermediation of the traditional banking firm. They will be caught by surprise, just as they were during the Great Recession, that their business models had grown obsolete—driven in part by their absolute resolutions in antiquated SOP axioms.

A Deeper Impact: It's all About FSBO's and Money

We need to internalize the principles that FSBO's (comprising a wide collection of segments supporting EVERY business in the world) is distinctive; bar no other industry, not even healthcare. Without money, its creation, its usage, its lending, its taxing, et al, the modern world simply stops. Bartering will not likely build fortunes.

Additionally, when mapped against the breadth of state and federal regulations, the negative perceptions held by many demographics, and the government involvement within every part of the supply chain, there are ravine disincentives for significant innovations and chasm-like barriers to systemic, sustainable change.

There are many things that can be reformed within this FSBO ecosystems; from multiple electronic currencies, to peer-to-peer payments and lending, to non-traditional asset classes to fundraising and employee compensation. It is a collection of industries that are designed to protect its consumers, borrowers and providers from hurting themselves. Let's face it, if a "money" supply is threatened, Maslow[xxix]'s premises become everyone for themselves—my survival, not yours. Like it or not, money runs the world.

Why we Know Disintermediation is Happening

Let's quickly look at one person, Ben Carson, United States Secretary of Housing and Urban Development (HUD), and his ideas regarding a staple of the varied middle class. If you review what Carson has talked about, he's not against the whole idea of replacing traditional benchmarks of mortgage lending with just the 30-year loan. If you review the implications of this singular idea and match that up against the treasury secretary who's got a vested interest in keeping the GSE's alive. Taken together, it is easy to conclude that there are better ways to accomplish housing and lending standardization.

The real issue, and seemingly what is at the heart of reform, has been that the amount of regulation really does not lend itself well to a complete disintermediation, especially on the retail banking side or for the small lending segment. On the mortgage side, you have an asset-backed security, which is really what they're talking about. That asset must be validated and verified, stretching out the entire process and relying on human analysis, the specifics of which have been largely untouched by time.

Whereas large asset-backed lending will be disintermediated, it won't take place for one to three years—thanks to fear, politics and keeping the financial supply chain status quo—as no one wants to be responsible for another recession or extensive government bailout due to sloppy or lax regulatory guidance for the asset-backed securities industry.

However, this is happening already in a different segment of the lending market. Just look at how car loans are made. Car loans are written in a matter of minutes, but they're an asset that is maybe one eighth or one fifth the size of a normal home loan for most of the consumer market. Auto[xxx] loans have risen nearly twofold in the last couple of years (as of this writing over $1.2 trillion and equal with student debt[xxxi]) across the country.

When it comes to tangible asset-backed loans that are larger, like a home, we are seeing higher rates of abandonment often due to an inspection or a validation of incomes by somebody along the chain trying to approve that loan (i.e., tougher underwriting standards to limit downstream repurchases). What this points to is a better method for qualifications on the upstream segment as buyers and sellers cannot live with rising sales that are continually being rejected downstream. We're talking about something that went from 2% to 4% since the start of the Great Recession.

That represents a significant amount of assets that are being rejected. We are not advocating for the early AVM initiatives and valuations, which historically demonstrated a lack of a true representation of a given property, but we are suggesting that there are sophisticatedly advanced efforts by firms whose executives will say, "I've got a better way to do this. We have a virtual way to do this, which we warrant, as it is far less variable, and achieves laser accuracy even under market duress."

Moreover, we can digitize the components of a home or tangible asset and come up with clear evaluations with greater granularity than historically possible. These changes allow homeowners or consumers to then conclude, "I don't need bankers. I can go seek out private funding. I can go and find a couple of different people to offer me the mortgage or seek out a hedge fund or corporation looking for a greater return form me, a qualified borrower."

This brings up a new market reality. Hedge funds and other asset-backed investment firms are now some of the largest holders of rental properties and mortgage loans in the U.S. since the Great Recession. These firms have gotten into the residential rental property business because of the void of lenders and the projection of greater returns for their investor portfolios. When you look at the valuations and the hundreds of billions that have been allocated, as an observer you could postulate that they may become the next generation of bankers, lending against properties for consumers, while reducing points and extraneous fees.

Again, disintermediation translates into taking out traditional processes, taking out traditional methods of doing business, and going directly to the consumer. That's what disintermediation is all about from an asset-backed lending standpoint, essentially, we are attempting to go directly to consumer, directly to that relationship. And if it's not a hedge fund or similar FSBO-like enterprise, then the argument can be made that people will get together and crowd source on Amazon and offer another option in lieu of traditional financial supply chains (see Chapter 9).

There are so many ways to make this happen. Some could even take a page out of the consumer or corporate lending to small businesses playbook and apply it to their startups. They would then be creating their own "virtual currencies" (these offerings are increasingly referred to as ICO—independent coin offerings) that are for now outside

SEC jurisdiction, and directly issuing their own "coins" to people as investment tools and saying, "If you give me $100 we will give you one coin" in lieu of stock certificates tied to the value of an asset or portfolio. Projections already have topped $180 billion from $17 billion, representing hundreds of percentages increase compared to the prior year[xxxii].

Those instruments may not have any monetary value or be traded on an exchange, but in theory they are no different than other fungible instruments gaining popularity. In fact, they may be "safer and sounder" in an asset-backed marketplace if they have a worthwhile property or future idea (i.e., personal fundraising) behind them. The initiating party can put that property or transaction into the market and individuals or enterprises with investment money can help complete the transaction without ever involving a bank. It has yet to be determined if these efforts are creating a shadow banking world outside government oversight, or if this is just part of an evolution facing traditional FSBO's.

This wouldn't be the first time a Wall Street firm has started offering banking services. We've seen it with some of the big Wall Street firms for years increasingly eager to lend money for mortgages and using their customer's asset portfolios as the collateral (others would rightfully argue that these same individuals and firms created the Great Recession).

Recently, in an Accenture global study, approximately 20 different countries and nearly thirty thousand people were surveyed across a range of FSBO topics. One of the results was that roughly one third of those responding said they would go and get a loan or get a retail banking product from Amazon, Google, or Facebook[xxxiii]—over a traditional, branded FSBO, if given the choice. These outcomes should be horrifying to banks.

They don't seem to be horrified yet. But they may be, if they eventually experience the future we have foreseen. Where are the biggest risks going to come from? Who is going to recognize them? Can an organization transform against the product hurricanes coming due to this rapid disintermediation?

A Deeper Reality: Planning is Not Delivery nor Acceptance

For enterprises seeking consistent movement beyond one-off marketing programs and good intentions, a comprehensive disintermediation strategy accounting for the Varian Rule[xxxiv] must navigate the following categories for sustainable success:

- Staffing, People and Culture
- Accounts and Products
- Customer Interactions, Fulfillment and Services
- Core Delivery and Integrations
- Prospecting and Marketing
- Investors, Suppliers and Boards
- Enterprise and Competitive Innovation
- Research, Sourcing and Due Diligence
- Credit, Instruments and Capital Markets
- Data and Process Management
- Regulatory Compliance
- Software Applications
- Technology
- Intelligence and Analytics

Regardless of whether the solutions are developed in-house or licensed from larger competitors (i.e., white-labeled), taking advantage of disruptive innovation requires a comprehensive, iterative plan-of-attack. Only by utilizing a holistic approach and analysis for technology adoption can enterprises be assured of repeatable successes across dissimilar, discrete digital advancements.

Marketing alone cannot make up for the void that is missing when impacts to business models and delivery methods go unheeded. Therefore, to address the questions posed, organizations need move beyond one-off ideals and towards the incorporation of abundant innovative disruption.

And even if the Rocket Mortgage offering fades (a gaggle of disintermediated innovations), the advancements and behavioral changes put forth will not—just as we would expect according to Varian's Rule. Who will offer the next iteration of Rocket Mortgage and what questions will be answered resulting in a 20% to 30% casualty rate among existing FSBO's in just four years?

Biggest Risks are the Most Immediate

The biggest, most immediate risk will be that FSBO's are cut out of the relationships they currently enjoy with customers. What will happen when they are not controlling that relationship anymore? What happens if financial data and/or customer contacts reside with partners or competitors rather than within the internal infrastructures FSBOs have built over the decades? Are we just left with the "financial plumbing" or some sort of a utility model across the FSBO supply chains?

If FSBO's become nothing more than a commodity or behind-the-scenes fulfiller within their markets, their business models will lead them to comprehensive ruin. As an example of this today, we see consumers seeking the best rates for their loan transactions, they hunt the lowest rate, the best terms, in some cases the fastest approval, as they feel empowered to shop from varied lenders in markets that were opaque and in some cases closed just five years ago.

Consumers are cutting out traditional FSBO's who don't feel compelled to respond to the consumer's new requirements for conducting financial services. When you incorporate that with sophisticated social media platforms and market makers buoyed by big data repositories and analysis (boasting countless statistics and information on the consumer or homeowner), the progressive FSBO's and startups are going to be the preferred stop for educated, information-hungry consumers.

We saw segments of this with Rocket Mortgage, the pioneer for "smart" home lending, but we're also expecting to witness rapid advancements from large data-advanced retailers (i.e., Amazon and its cashier-free stores) as leader-driven enterprises try to figure out how they can increase profits, boost margins and gain greater loyalty from bespoke buyers.

The first challenge for most FSBO's that do not own the consumer relationship will be to somehow counter their business becoming a commodity based operation or set of offerings. As the ability to disintermediate existing products, services, and fulfillment moves along the FSBO supply chains and impacts the relationships between partners, (e.g., who are doing mortgage appraisals or mortgage insurance and all those other things), these once valued products and services will be retired due to lack of interest as FSBO's recognize that they no longer can influence that channel or product suite. The channel is no longer going to be driven by the institution's historical relationships with the consumers. Whoever is providing the delivery platform or mechanism will now control product and service consumption advances, directions, and offerings.

FSBO's used to be engaged in a comprehensive, relationship-driven business. Financial services used to operate under a model in which employees went in, talked to people, shook their hands, and ensured that the consumers understood the discrete differences when analyzing competitive products. That has all been transformed through technology, social media and the whole digitization effort.

Technology is now the initial and lasting touchpoint and consumers like it that way, due to their high degree of skepticism, distrust and outright animosity aimed at "bankers." All of this continues to be the demise of "professional bankers." If FSBO's dig their heels in, remaining tied to the belief that they are a necessary component of every consumers' financial society, while constrained by regulations imposed forensically due to the actions of bad actors, their fortunes will continue to be dim.

Even with those hoping for and anticipating new U.S. government administration adjustments, removing constraints or making oversight and regulatory compliance simpler, the impacts to business models may accelerate as new entrants take advantage of FSBO cultures who are thinking the worst is behind them. Bottom line, the removal of market regulatory constraints is a double-edged sword for banks. It helps them and it hurts them because it opens new vulnerabilities and new market capabilities.

Talking about FSBO business as a relationship business is going to sound obvious to a lot of people working in other industries. It's not really that intuitive here in banking, where for the longest time the customer was considered a "work input", and the real relationship was with somebody in the secondary market. This was especially true in mortgage lending, but also in some of the other bank products where non-deposit money fueled consumer lending.

FSBO leadership may not realize how serious is the risk that they could lose the client relationship completely, but we promise you it is. The only question is how serious, and that will vary by institution. Back during the Cold War, nuclear scientists used a clock to tell us how close we were to doomsday. The hands were poised at 11:59 PM.— That's how close we were to doomsday.

Today, we believe the current clock for FSBO's is set at 11:57. For many of them, the clock will soon strike midnight. We expect it come as a complete surprise to FSBO leaders as they seem to be unaware that the building blocks in the form of the compartments of applications, infrastructures, orchestration roles, and architectures, which have been falling into place for some time. Soon, some FSBO, FinTech or startup will begin to assemble these blocks in a certain way, which we expect to happen at Amazon and Google. We believe that progressive companies which understand how to create and leverage a customer profile, have the big data to create bespoke solutions, and use these high-value expertise's to obtain first-mover advantage with long-term adaptability—and then the clock will strike midnight.

These tech-savvy, consumer-oriented enterprises are going to have best-offer informational solution sets available at their fingertips. That will be an unwelcome surprise for many of the FSBO's, who since the U.S. election of 2016, have been waiting for a regulatory reprieve, which would take the industry back to the way it once was.

If we are right about all of this, then the ability to provide rudimentary banking services as well as the ability to lend money to different market segments via diverse financial instruments, which don't look like traditional bank products may become much easier.

Currently, the top five banks (Big-5) control nearly 50% of all consumer finances in this country. That is a great deal of risk, by simple virtue of the concentration of assets. But these institutions face a great many other risks, to their brands and reputations thanks to countless stories carried in the media and on consumer news feeds, either real or fake, all of which serves to drive consumers to someone—anyone—new.

Larger FSBOs that bet on the repeal of current laws and a future in which regulation was less burdensome, are likely to be disappointed as we do not see the controlling party scrapping Dodd-Frank or major parts of financial regulatory oversight. The implication is that these firms may have to increase their available capital levels to the tune of $500 billion to $1 trillion dollars[xxxv] just to weather the storm.

Meanwhile, the financial services industry continues to evolve around the major leader banks. We are already seeing this with new startup fundraising. New firms don't even go through traditional VC (venture capitalist) channels for funding rounds today, to say nothing of traditional FSBO's.

One company was planning to raise $5 million for a FinTech startup. Executives putting the deal together expected it to take 30 days. They were able to meet their cash needs in just 58 seconds using their own "currency" (i.e., an ICO[xxxvi]) and going directly to the "investor" via an auction website, walking a fine line just this side of requiring SEC approval. Investors bid for it and saw direct value, meeting the company's need for a rapid infusion of capital, and eliminated the advisory services and fees common during traditional FSBO fundraising stages in the process. Where are FSBO's or VC's going to make their money with this new generation of business enterprises?

Why wouldn't the same thing eventually happen in the mortgage lending world? Or for any other traditional banking product? We expect it to happen and the hold traditional FSBO supply chains have on business will go away. Regulators, who have traditionally been no friend to bankers, are now the only ones standing between the status quo and the comprehensive disintermediation of products and services we have described. Bankers that have been crying for relief from their regulators should be careful what they wish for.

If, as an FSBO, a company hopes to mitigate the risk of disintermediation, a series of action provoking questions should be asked: What us the timeline for a transformation effort like this? How much time do we have? Is this a five-year project? Two years? 18 months? How much time as an enterprise are we willing to allocate when opportunity or adversity brings us into the market to raise money or secure a loan? What is our model as a business? What really generates money? Can we really understand these boundaries and break them down into the parts (compartments), that are or can be disintermediated?

What happens if the U.S. President suddenly says banking needs to be cheaper and more widely available to even those who are underbanked or unbanked (and, we see how he tweets about Amazon, an "American success story")? The administration can jump into social media and create a strong-storm of 140-character streams sending their followers and constituents into a frenzy against those "evil bankers." We have seen this happen to other industries already.

In essence, "I don't like what they have to say[xxxvii],"

once tweeted becomes a scenario that reinforces existing animosity and cements it into the buying and consumption patterns of all customers—whatever the reality may be. Moreover, if something like that happens, and the FSBO has already invested in the financial tech startups (shared or captive), there are many things that could happen as a new financial supply chain sets the stage for a commoditization of FSBO offerings (see Chapters 4 and 9).

A Deeper Advancement: Cybercurrencies the Next Frontier

In a finance world seemingly obsessed with everything digital—consumers, statements, mobility, payments, compliance—the one thing that is now generating growing enthusiasm (or hype) is currency. Currency in this context refers to a multitude of labels, some more subtle than others. And when you listen to experts or read articles, it quickly becomes a discussion of mathematics and technology sophistication that can be well, a little mind-numbing.

Some labels refer to them as cryptocurrencies (e.g., bitcoin, nearly a thousand others), digital payments (this could mean anything paid electronically), virtual currencies (i.e., not intended for "common" payments), and of course, still emerging is a central banker's vision of a digital sovereign currency (e.g., Bank of England). Moreover, further distinctions are made if they are peer-to-peer settled (i.e., part of the "sharing economy") or centrally cleared. Supplementary attributes can be attached regarding convertibility, closed versus open, regulated versus unregulated, protections and even exchange rates.

With cyber protection, fraud detection, money laundering identification, consumer protection, and technology sophistication increasing, merchants, consumers, regulatory authorities and even banks are slowly entertaining wider adoption of, for lack of a better identifier for all the above differentiations, electronic currencies (EC). Therefore, pardon our glossing of distinctions to portray the impacts challenging the organization, regulations, business models, business practices and solution adoption in a digitization of everything (DOE) era. The emergence of EC's also may offer a distinctive set of offerings to stymie disintermediation, while temporarily slowing commoditization of offerings.

In total, EC's represents a sophisticated and emerging set of global solutions that never fail to capture the interest of bankers, media personal and most recently conference organizers. However, while the complexity and sophistication of currencies in these digital formats demand equally high-tech discussions, the repercussions of their incorporation into the more global financial supply chains are beyond those open-source collaborations or digital ledger advancements (e.g., blockchain) making headlines in press releases.

Whereas understanding current and future advancements in EC's is imperative to successful implementation and ongoing adaptation (as robustness and features increase with time), treatment of their impacts to existing FSBO processes and operations requires a comprehensive impact analysis across existing products, services, alliances, outsourcing arrangements and infrastructures. EC's have the potential to be a catalyst for a fundamental shift in how digitization is applied to FSBO's—a move from one-off projects to comprehensive transformations.

A Disintermediated Reality—The Rise of FinTech

The rise of FinTech was aided to a great degree by the industry's misplaced hopes that regulation and oversight would be stepped back. This put traditional banking business in a waiting pattern. Meanwhile, the upstarts capitalized on the legislative fragmentation created by state and local governments in their efforts to regulate within their own bubbles. They found ways to move into the industry.

We expect to see technology-based Fintech firms use this to their advantage against traditional financial services firms, especially those that are federally chartered. A state chartered Fintech firm would enjoy various capabilities to raise funds and offer those funds to homeowners at a local level. Then, through a clever partnership, they could offer those firms to a much larger market, threatening national banks. Imagine such a firm working through an Amazon or some other portal. The new institution could maintain its local presence, but also gain national visibility brought on by the partnership and the resulting expanded supply chain capability.

That is why FSBO's are going to quickly find themselves competing in a market during radical transformation. If they are smart, they will get ahead of it. This metamorphosis may take five years to get completely in place, but the smart FSBO's will understand by the time the next midterms arrives and the "fast repeal" which was promised does not appear, means that are looking at a completely different competitive landscape.

Strong job growth[xxxviii], steady consumer sentiment and more factory orders belie a movement by consumers for a redistribution of financial controls; beyond government and beyond name-brand FSBO monikers. Buried within the populist's messages taking place in the U.S. political maelstroms we call "nominating processes," are fundamental FSBO shifts. These shifts are major factors in the rise of FinTech solutions and even FinTech and retailers as financial institutions. If in doubt, try mapping them to digital technology shifts and you might be surprised by the associations.

A Deeper Dispersion: It's More Than Words and Slogans

The desire for disintermediated innovation within many leadership circles is an aspiration to "be competitive". What they are really saying is that, "we want to improve efficiencies, improve profitability across products, services and demographic targets, and we need to grow beyond M&A (i.e., expand organic offerings)".

To focus enterprises, the umbrella of expanded digital banking capabilities along with its vast digitization of data has emerged as a critical strategy and a core competency from community banks to credit unions to systemically important enterprises.

Additionally, many financial enterprises have a high fixed cost of operating leaving limited capabilities for preparing for an enhanced Varian Rule outcome. Though many institutions are touting their internal IT startups, patents and advancements in transaction speed, the reality is many are still leveraging legacy systems and using government dictated processes, which add to costs and limit consumer responsiveness. As we say in the Midwest, "it is like putting lipstick on a pig."

However, planning for digital change often requires a substantial modification in business approaches or even models of operation due to accelerated advancements with a solution or technology grouping (e.g., smart phones).

An example of how solutions once utilized by only those with the largest capital budgets have made themselves available to the mainstream FSBO can be seen within Amazon's Web Services (AWS) and their most recent addition, which migrates in house databases to the cloud with no downtime or expensive porting software—and they are even able to switch to different database engines. A decade after it rose to a consistent business priority to drive down fixed costs, cloud computing continues to expand and is now available to every enterprise (per the Varian Rule).

However, what is strikingly sad is that when it comes to preparing for innovative disruptions, the specifics within many enterprise plans penned by expensive advisors are missing or woefully inadequate for what is forthcoming. Since the 1980's "watch words" such as optimization, performance, transformation, agility, efficiency, and so on have appeared in nearly every communication in what enterprises will be doing to leverage (continuous) disintermediated innovation. Yet for nearly four decades, FSBO leadership is caught-off-guard when technological advancements outpace their ability to predict outcomes. The failure of execution resides in a fixation to narrowly focus on one or two advancements—like we are doing today with distributed ledgers[xxxix] and mobile apps. The reality is that the idea of the Varian Rule is being fast-tracked as multiple advancements converge—machine intelligence, investment robots, big data, sensor advancements, security, privacy, VR (virtual reality)—and embraced by a consumer base that distrusts and eschews traditional financial solutions.

The Ultimate Impact on the Traditional FSBO

If traditional FSBO offerings are disintermediated, and consumers have a good reason to believe that they no longer need traditional banks, which given what we've already discussed about their general dislike of the industry and its participants seems likely, how can FSBO's survive? Where should they be looking for impacts to their operating models? Will this generate further consolidation from just over 6,000 FDIC institutions to far fewer?

If we look at the foundations of what this means, those risks are serious, and there's a fundamental shift going on. We see that, not so much because of technology, but the fact that the U.S. real estate market and the world markets have changed. Our existing processes for how we select and incorporate solutions and retire technologies have all become antiquated.

Again, the question is not whether these risks exist, but how serious they are. Are we talking about potentially game-changing transitions that will put a large percentage out of business? We think it's a severe issue and FSBO's must consider these risks carefully as they plan not only for current budgets, but future years' budgets.

Many FSBOs will continue to lock up large budgets that are tied to playing the regulatory compliance game. This is short sighted. Not because we think regulation will go away. We believe it will become more intense as Fintech firms penetrate the market and begin to wreak havoc with consumers. And while there are certain budget items that cannot be ignored, such as protecting consumer privacy, data security, and compartmentalization assembly, it will make no sense for FSBOs to become very safe, very compliant providers of nothing. If they continue to allocate the lion's share of their budgets to compliance instead of investing in the future products and services consumers demand, to say nothing of the delivery methods they require, this will surely be the result.

The shifting business and operating FSBO models indicate that the industry must find a way to simplify its invention and delivery times. FSBO's must cut the market lag when competitive pressures move beyond existing capabilities, even if it means forming their own partnerships with leading consumer-facing organizations to keep those relationships with desired consumers. This will become the key issue and critical competency for many.

Failure will bring with it serious consequences, which will be quite apparent to leadership as market and competitive events unfold in front of them. But many will miss the root consequences as the event will be wrapped in failed product launches or declining margins when the real causes are eating away at the models of business operation and customer delivery.

It is a high probability that negative FSBO events will play out in parallel as the missed implications corresponding to product and service rationale show up in quarterly results and investor displeasure. A fundamental result for many FSBO leaders is that if their BAU models are failing (precisely due to a dogmatic adherence to the models), consumers will exit these laggard organizations due to their inflexibility and poor offerings.

Clinging to their current practices out of fear of the highly regulated environment or standing, in the hope that a less stringent environment will present itself, will lead to failure. It's not about your compliance budget. It's about your business operating model. When that fails, the company fails.

Clinging to current practices out of the mistaken believe that the company has brand leadership, a strong reputation, and a great many consumer accounts will also lead to failure, but do so more quickly. As account holders begin to bleed off, consumer trust and loyalty erodes—along with more and more of the institution's operating capital. The FSBO may feel it can weather the loss of markets and contact points, but we expect that they will eventually realize that the losses are unrecoverable.

And there are other trends that will exacerbate these issues, including increasing M&A actions, liquidations[xl], privatization efforts, and other high-end partnerships dealing with emerging technologies. Another issue relates to foreign workforces, which has a significant impact on FSBO technology initiatives. Many banks are already taking advantage of outsourcing[xli] components in their back offices. How will they deal with these in a future in which H1B programs benefit fewer companies or more? What happens if these are restricted or made only available to those with "scarce skill sets?" How will that impact the FSBO workforce, and its ability to innovate, in a quickly changing competitive market?

Does that mean the H1B's really go to innovative technology companies like Google and Amazon which are generating jobs for and within the country versus H1B's that are brought into this country simply for labor arbitrage? Outsourcers are creating jobs, but they are often foreign jobs, which are reporting back to a labor exporter (arbitrage) country. The result of this perpetuation is that there will be further commoditization of these collective trillion-dollar back-offices which banks have built across product and services.

It's going to be hard to change these dogmatic processes, and it leads to another reason that many believe FinTech will replace FSBO's, as these new firms don't operate under the same mindset. For many of these startups, this leadership is young and not mentally constrained by prior norms or operating processes. They are looking at something different, new, exciting, and sexy that not only fits their lifestyle, but other consumers who are just like them. They may not have a banking background, but they see a better way to do financial services without the constraints and archaic rules of the traditional process.

Moreover, these "upstarts" are well capitalized. They don't care about disrupting things. They have no fear of failure. They don't know what can't be done and it doesn't scare them. They're loaded with attorneys thanks to full coffers (some via ICO's) coupled with the need to protect unique ideas and innovations. They don't have to worry about monetizing their models yet. It should be terrifying to traditional bankers, but to many have a false idea of customer loyalty leading them to believe they have market lock-in and systematic importance.

A Deeper Intrusion: When Pressed, FSBO's Play Politics and "Big Brother"

For financial services firms, dealing with loan margin declines, declining traditional product mixes and an increasingly mobile, irregular series of consumer behaviors, the achievement of transformation realization (i.e., beyond slogans and words—booked results) cannot be unprotected and slain by political selfishness and manipulation, especially when FSBO markets are being holistically disintermediated.

Individuals who utilize politics and big brother against progressive disintermediation are very limited in their capabilities and value-add when it comes to the emerging FSBO's. They either fail or refuse to process the fundamental or critical shifts taking place around them to satisfy current and future customers.

An example future case study along these themes might be the top U.S. FSBO's roll-out of an employee surveillance and compliance systems that flags "potential abuses by employees over hundreds of digital data points" before wrongdoing is even done (i.e., sanctioning intentions)—it's a big data application akin to the **Minority Report** movie. It's a strategy to reduce legal exposures, but politics of its realization will likely result in chaos during its roll-out and evolution. It might be a case of being TBTF becomes TBTS (too big to succeed).

As we can extrapolate in this process of gamesmanship, the politics can warp and twist the achievement process resulting in PU's (profit underruns), staffing defections or worse a sickness across the customers (i.e., "…a House on fire cannot survive[xlii]"). Repeatedly, these motivated leaders are so driven to set benchmarks of excellence that they get caught up in the competition of strategy implementation (win-lose) and evolution (mutant strains) resulting in hidden agendas, retribution, and fear—along with program and initiative failure.

The internal operating dynamics become drama filled and tense, and the game spreads to all corners of the enterprise. It is a realization that if your peers are maneuvering, then if you don't, the consensus is you will publicly fail.

So how to you recognize the "Game of Maneuvers"? A snippet of the tactics includes:

- Do as I say, not as I do.
- Changes within existing initiatives which encompass new strategies, usually with an intent to obfuscate.
- The shifting of powerbrokers to protect or weaken positions.
- Follow the money, budgets and where efforts are being expended to determine trustworthiness and real intent.
- Beware multi-party discussions as these are often less about clarity and more about blocking.
- Trust me—ah, no, not without verification and validation.
- Expensive "plumbing overhauls", while improving

the infrastructure and processes, lead to delays and shifts in control and accountability.

As presented, disintermediation is sometimes unpredictable, but it can be managed and measured the science of results. However, science and results require people and organizations working in harmony towards common goals and objectives sometimes with great personal sacrifice. Consequently, dealing with continual disintermediation is far more than a program and critically dependent upon people and the muting of their organizational politics.

Competitive and Consumer Assaults

Disintermediation caused by established and upcoming enterprises is no longer just about FinTech—it's created by cross-purpose institutions spread across hundreds of industry segments. The new entrepreneurs generate apps that are easy for people use. They can do this because they study consumers and know what apps they already utilize daily. They are experts at mining online information and they share it freely with other likeminded entrepreneurs. They know how to tap into and create profiles that become indicators of the trustworthiness of qualified lenders, and they use this information to secure consumer trust and acceptance. They know how to get into the white spaces, find out what is in demand by consumers, and how to profit from it. These are advanced functions bankers never had to do and have not learned to do as they have always believed they were immune to atypical, competitive disintermediation often due to regulation barriers.

For example, FSBO's fortunes were built on paper forms. They sent people in search of information, even though it existed elsewhere within their organizations boundaries already. Their process was built around the customer giving them information in exchange for a result. But the new consumer-facing apps get their consumer information without having to burden the consumer with providing it.

The attitude within these market-disrupting enterprises is completely different too. These "upstarts" are used to living and dying by creating apps (some knowing were transitory in design) that users want to engage. However, bankers have traditionally lived by delivering what they perceived consumers to need (as opposed to presuming that they need consumers). It's a different approach to the business, which will continue to have a big impact on topline numbers.

Many of these "convert" consumers are beginning to say, "I don't need a bank to tell me what I can or should do. I don't need a bank to necessarily provide me non-appealing products because the government says I need 'safety and soundness.'" They're saying, "I'm willing to take a risk with a new financial provider. These new FSBO-like institutions may not value all my information to the same degree that traditional FSBO's have, but if they can guard my personal information, my identity, my financial life, isn't that 'safety and soundness' in a different way?"

If a company were willing to offer traditional FSBO consumers the ability to have finances "their way," then why wouldn't this be good for consumers, markets, competition, and global movement of money? The rub will always come (e.g., Bitcoin[xliii], various cybercurrencies, and ICO's), with an acceptance of a new model for finance that pushes the boundaries of what regulators and risk-takers thought they were doing[xliv].

It is worth noting that Bitcoin (along with other cybercurrencies) is one of the greatest (or worst) assets classes we've seen in the past several years—it moves to the extremes of biggest loser and winner over and over[xlv]. It went up 30%, then it's down 40%. It went down 40% in three days and gained it all back plus some a few days later. Some individuals look at that and say, "Wow, this is the new model for commerce!" People seem to love and swear by virtual currency, ICO[xlvi], or cybercurrency, depending upon how you want to classify it. What's does it have? It has blockchain[xlvii], it has a distributed ledger, and a host of other financially advanced technologies that make it useful, profitable, ubiquitous, and dare we say, sexy. It's nontraditional, but since 2011, its components are now seemingly everywhere in the traditional FSBO operations.

There are several reasons why banks are so eager to not only embrace Bitcoin-like principles, but mostly it centers around their need to develop and patent their own solutions as a method to lure consumers through differentiation—and ensure future lock-in of consumers (and some institutions are banding together, e.g. R3[xlviii]). FSBO's feel they can address the desires (not needs) of the financial consumer, and then they can provide the safeness and soundness of transactions as they have always done in the past.

However, there are many people challenging FSBO mainstreaming of emerging financial products and offering consumers a less ridged and transportable (i.e., not locked in) set of offerings. We think that's where digitization and that whole unintended consequences will further propel advances in supply chains but also rapidly expand the cycle[xlix] of disintermediation and commoditization of new solutions.

There is a common misconception that when FSBO's launched holistic digitization, around 15 to 20 years ago, that these top-to-bottom programs were viewed as a "one and done" initiative. The axiom that moving forward the breadth and depth of electronic records, their correlation between internal and external sources, broadening of data into vast repositories spread outside industry pillars, and a consumer pushing FSBO's to accept a lack of control over traditional channels, was never contemplated.

This failure to anticipate, let alone embrace, on-going advancements has materially contributed to an inability to resonate with non-Baby Boomer demographics. Moreover, these pervasive failures have provided the genesis for disintermediation as non-FSBO industries sprang up and encroached into the products and services once delivered by brick and mortar locations—and shunned increasingly in favor of customer centric solutions (the way they want to be served).

From the consumer's perspective, are they going to be better off if banks go away? We don't think banks will ever go away. Maybe they will go away in 50 years, but we believe over the next 5, 10, or 15 years they're going to look very different. Furthermore, we believe that it will benefit the consumer across an array of financial solutions.

In general, anytime we disintermediate a market, we get greater value for the consumer. This usually removes non-value-added expenses, which for instance add extra basis points onto a loan just because all they did was push a couple pieces of paper. Cutting out these non-essential middlemen allows for great service consolidation. This allows FSBOs to go directly to their customers, offer them an evaluation expert who can then help them get the loan they want. Those types of services can be replaced. They can be augmented in different way as part of a disintermediated channel, which in theory will always benefit the consumer.

Anytime things are made more transparent and less complicated, we usually see a greater benefit to the buyer, and that greater benefit can also be seen in the secondary markets. For example, if you want to examine the increased value for a lending product because we have fewer people in between adding superfluous information to what "we're doing", an investor can transparently assess that asset (class) and how that asset is bundled into a private instrument[1] that can be sold into the secondary markets. This results in more accurate risk measurement and open pricing. We know, as investors, what's in those loans and whether they are covered bonds or whether they be new mortgage backed securities. In this case, disintermediation will speed and promote transparency and due diligence abilities.

FSBO's will be the catalysts, which in theory, then draw more capital thereby benefiting the consumer more. It's like a priming of a water pump. In this case, as disintermediation happens, it really is a priming of the mortgage pump. This is happening now, which makes it all the riskier for banks that don't pay attention.

In an increasingly disintermediated financial world, FSBO's must pay attention. FSBO's must partner. FSBO's must understand that the models they used for decades are no longer the "best" model, because they don't control the relationship between the consumer and the offering. And no amount of regulations or deregulation (see Chapter 5), will recreate the prior models of prosperity.

Illustrative Challenges / Opportunities

		Controls and Implications (The 4-V's)			
Cate-gory	**Illustrative Challenges / Opportunities**	**Volume/ Rate**	**Variety /Adapt**	**Velocity/ Freq.**	**Veracity/ Accuracy**
1	Power shifts towards consumers and their preferred channels provide the impetus for FSBO disintermediation of product and services.	↑	↑	↑	↑
1	Data availability and complexity provide the pathways for competitive pressures for gaining attention and trust of traditional FSBO customers.	↑	↑	↑	↑

Cate-gory	Illustrative Challenges / Opportunities	Controls and Implications (The 4-V's) Volume/ Rate	Variety /Adapt	Velocity/ Freq.	Veracity/ Accuracy
1	Rapid investments in FinTech segments has created best-in-class performers outside traditional FSBO's resulting in shifts of asset placement and with transactional based purchases.	↑	↑	↑	O
1	A failure of strategy highlighted by incremental FSBO innovation has left enterprises with commoditized products and services that are low-margin, and price sensitive.	↑	↑	↑	O
2	While legal and regulatory structures can inhibit growth, it also retards normal disintermediation resulting in an ebb and flow underneath political pressures.	↑	↑	O	↑
2	FSBO cultures are not accepting nor are they architected for disintermediation of high margin products and services with the spread of FinTech, AI, VR, and robo-investment.	↑	↑	O	↑

Cate-gory	Illustrative Challenges / Opportunities	Controls and Implications (The 4-V's) Volume/ Rate	Variety /Adapt	Velocity/ Freq.	Veracity/ Accuracy
3	Enterprise focus on task-based cycles of create, develop and deploy may aid with large scale efficiency, but they fail to grasp the demands of adaptive and agile cycles necessary for consumer driven innovation behaviors.	O	↑	O	↑
3	Operation performance measurements and accountability are failing as they struggle to accurately reflect advancements and the upstream and downstream implications thereby leading to false-positive decisions and rewards.	O	↑	O	↑

↑ – Increasing non-linearly, unpredictable behaviors
O– Increasing but predictable

NOTES

"This is us."

Image ©Shutterstock, Inc.

4

The Risk of Commoditization

"*Banking is necessary. Banks are not.*" – **Wells Fargo's** 2004 annual report.

When it comes to commoditization[li], FSBO leaders don't seem to think it's even possible for their institutions—except within old product lines encased by "innovative" front-ends. Their beliefs have been reinforced when they see new financial products being offered across existing and emerging devices and channels. "Well there's new 'stuff' all the time, how can what we do become a commodity if we're creating new 'stuff'?"

The real risks for FSBO's is a shifting of how people use financial products, and how they view the services that are offered. That includes customer service down to how they charge fees they use for accounts including the funneling approaches employed to influence consumer usage[lii].

If you look at the changing demographics, much of the principled marketing strategies are adaptations of Baby Boomer rationale and implications migrated to the younger demographic segments. Within the United States, brackets across the Baby Boomer market segmentations are being forced to start the extraction of money from retirement funds that have been built up over decades. The build-up of Baby Boomer assets, and its multi-decade transference to younger generations (e.g., Gen X's, Y's, and now the

Millennials), is currently ushering in not just asset shifts, but permanent, pervasive and bespoke (i.e., custom tailored to individuals) behavioral shifts.

Post-Baby Boomer generations, and the multitude of sub-segments, bank very differently than the branch-centric delivery models still clung to as viable by traditional FSBO's. These emerging and constantly transforming market "non-branch" defined segments have different goals and different needs—society benefits, value in free time, focus on personal, animosity towards FSBO's, and the importance they represent. If you're one of the big banks and you're one of the Big Five that controls over 45% of the market, you may look at these shifts and say, "Well, we are the biggest. We have large value assets, and we control a significant portions of the $15 trillion dollars (USD) in consumer finance."

An example of why commoditization is accelerating can be seen across the challenges quickly repeating themselves within and between numerous micro-segmented demographics within the Millennials. This large grouping says, "We want to send or lend money peer-to-peer and transfer the liquid asset without having to pay a bank to deal with settlement." They don't particularly care who can provide this service, but in the absence of availability from traditional FSBO's, they will seek out those who can meet their demands.

FSBO's respond with of all things, "This is a highly-regulated environment (financial services) that we track and settle especially under the rules of know thy customer (KYC), so we have to charge a high fee to cover our costs." However, these consumers understand that fee-based income is how a bank makes their money for retail consumers, and they believe the system is rigged to benefit the FSBO's who they hold with such disdain. There is an American Bankers Association (ABA) infographic[liii] that was released in 2016 showing that 71% of consumers state they'd rather go to the dentist than listen to what a bank says. So, the market and their consumers will continue to speak with their actions rather than being scolded for seeking out value.

If you start to put that in perspective the obvious question is, "What is going on here?" The Baby Boomers grew up with the belief that FSBO's provide value-added products and services (unlike the generation before who witnessed the hardship imposed by bankers during the Great Depression and a World War). They knew their banker and conducted all business via the branches (some even came to them like doctors of a bygone age). The post-Baby Boomer generations have alternatives brought forth by technology and virtual banking that destroys the axioms of dogmatic financial behaviors, and puts the power into the hands of the consumers.

Traditional FSBO products and services have been commoditized and in some cases destroyed FSBO's due to a false belief in "If I say no (to you the consumer), you will listen". It's much like a parent telling a grown child not to do something—they don't have to follow what is being presented as fact, and can look for options that they want beyond the "take it or leave it" big FSBO industry guidance.

A Deeper Message: Look Beyond the Hype and Also-Ran's

What appears to be accelerating across the FSBO marketing messages are "cooler" interfaces, "why wait" delivery, "your lifestyle" features, and of course my personal favorite "faster everything". Whenever we hear these themes we picture competitors performing the Keystone Cops[liv] routines jumping in and out of their cars seeking to catch the other.

Globally, as interconnected society's move from mobile interfaces (e.g., PC's and laptops) to smart-device interconnectivity (i.e., small, personal or wearable intelligent artifacts connected to ubiquitous Wi-Fi) fueled by demographic, age and technology advancements, there are fundamental and permanent shifts taking place. For a visual of regarding various FSBO pressures, the graphic below shows an illustration for an institution and its degree of efficacy for the categories covered.

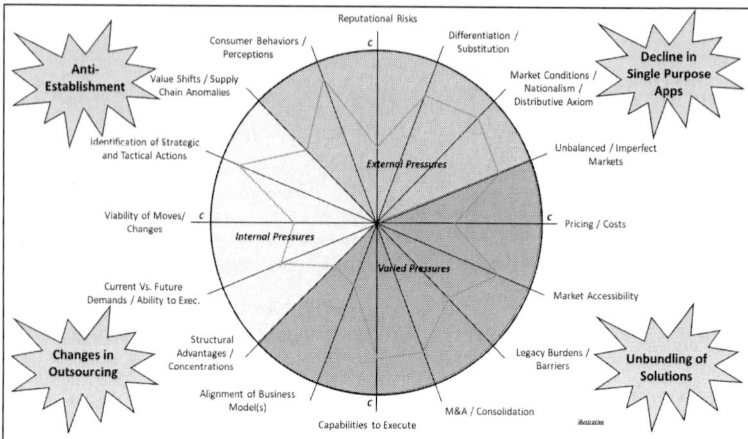

As an example, payments and clearing (pressures within the domestic and global markets) represent a concentrated focus as smart apps displace the need for cash while accommodating the growing demand for P2P transfers. Yet, if we look beyond the hype, banks or tech firms are trying to propagate traditional principles by simply using a new wrapper.

If they don't, impacts to fee-based income, market volumes and financial supply chain importance may diminish. Therefore, while banks believe they are "ensuring their future" they are alienating themselves from their consumer base, and setting themselves up for consumer advocacy blowbacks.

Marketing is Not Customer Innovation

We're seeing that branches really are a dying dinosaur—as they are commodities staffed with personnel who can only offer what's in the FSBO's BAU playbook. Back in the 1990's we used to say that the branches were dead. Then we saw a resurgence again when the markets increased and people wanted to trade (in equities and bonds). These active assets grew as the real and perceived value of the economy surged up to the middle of 2006.

As the stock markets deflated, we witnessed mortgage explosions due to the speculative flipping of houses, HELOC growth, and the seemingly never-ending increases in physical asset values tied to homes and property. The challenge quickly became that all these financial expansions stopped, the FSBO's infrastructure, personnel, systems, intelligence systems, decision making were all counter phase to the needs spiraling downward around their business models. The ensuing chaos, unpreparedness and unwinding witnessed market segments ranging from the most-trusting (i.e., Baby Boomer) to the least trusting and loyal (i.e., Millennials), which firmly believe that, "I really don't trust you guys."

So, today much of the focus in on Millennials[lv] and Baby Boomer market segmentations. This focus in often do to the fact that the generations in between, the Generation X's and Y's are financial caught between their children and their parents. They're having to support all the extended family financial obligations, and are not viewed by FSBO as important as the up-and-comers who have captivated the profit imagination of leaders seeking to reinvent their brand.

These generations are not necessarily so worried about banking products as they worry about just getting through. You look at the Millennials which will be representing 45% of the workforce by 2020, and FSBO's believe that to succeed they need a share of 50 percent of the 78 million people in this segment who right now say, "We don't really want to deal with banks. We don't think they offer anything unique."

You've got financial startups like Simple Bank who offers free checking, and has partnered with larger banks to create a larger, whitelabeled brand for their infrastructure needs. These startups make critically important, within their cultures, data which has the value, the apps for consumer experiences, and creating single banking delivery capability that is intuitive and rules-based—as the user desires. You can't just develop mobile app and think you're going to make a market where the competition cannot reach. Mobile apps and their common markets are commodities—those are things that everybody expects you to have. The ability to understand the consumer, anticipate their actions, and provide value-added services, even if you don't control them, are what makes market leaders today for a consumer.

When you put all the perspective of how the demographics and markets change, the commoditization is becoming very apparent within and across established FSBO ecosystems. As the branches are closed and we start decreasing 5, 10, 15% of the brick and mortars on a year-to-year measure, what remains is a different feel and purpose for the ones that are left (and there will for the next two decades), a need for physical branches if nothing else to meet products requiring a customer's' regulatory presence.

The FSBO branch is fast becoming highly interactive, and at the same time, delivering outstanding customer service—but will consumers continue to pay extra for products and services they can obtain from the comfort of their phone? Physical touchpoints to the consumer are solutions traditional bankers are familiar with as they have spent since the 1990's billions developing their customer solutions (or what they believe the customer should have). Those solutions, those infrastructures really have very little to do with the future of the markets as the most sought-after consumer segments will never use a branch (your co-author Mark has not been in a bank branch for over 13 years).

That's where and why you're beginning to see the commoditization across FSBO's. Couple this decay across "branch" ideals and solutions with declining profits and you're also going to see an acceleration in M&A over the next decade as banks are either going to acquire advanced solutions or they're going to start shedding them (i.e., outside their competencies to reinvent). We think you saw that recently where one of the top five FSBO's said, "We really don't want to be in the mortgage lending business. We're being successful without it. We don't want to get into some of these traditional products because we can't add any value. We can't differentiate ourselves and add to our brand where people want to buy."

Commoditization surfaces when competitive products and services are functionally the same leading to a void of consumer enthusiasm and a perception that the provider's solution does not fit into the lifestyle of how financial products are delivered. Even if there are a few different features, in the consumer's adoption habits, your FSBO solution represents commoditization—a non-value-add. Financial consumers aren't going to pay for what they can get easier and for free (or limited fees) and you're seeing that in banking today. Stock trading is an example of what typically cost the consumer $100 to $200 per 100 shares trade now cost as little as $4 all within a nine-year time frame.

A Deeper Option: Outsourcing in Layers

FSBO's cannot continue to operation BAU by attempting to move forward, while having their focus on the rearview mirror (i.e., doing things similar to what they have done for years). FSBO's need to adopt the new layered approaches to outsourcing (e.g., orchestrated, networked) and take a centrist view of technology and process architectures, while partnering with many likely outsource providers both domestically and internationally.

Additionally, you need to find partners (i.e., traditional outsourcers, vendors, consultants) who can provide not only a sounding board for your directions, but who are willing to provide unbiased, external "...did you think of ..."—even if it is not what they currently offer. Finally, accept that market movements and consumer trends will not be measured in years moving forward—but in months.

Those of us who started the ITO and BPO efforts over two decades ago are shaking our heads in wonder at those who vehemently describe the state-of-the-industry with models that were already antiquated nearly 10 years past. For those FSBO's leaders who are seeking a profitable way forward, the idea of layered outsourcing should hold immense interest and potential—as it provides means to address key regulatory areas of protection, transparency, security, and adaption (e.g., organization stress conditions).

FSBO's No Longer Dominate the Channels

You can look at the wealth scandals with fee based income, the internal cultures which promoted fraud, and even the industry leaders who turned a blind eye, customers and would-be consumer are saying, "Whoa, whoa, whoa, you guys are all the same. You all have the same product. Why am I paying for this? Why are you allowed to behave badly—I don't want anything to do with you!" Bad actors (and it seemingly never ends) just reinforce the animosity so many cross-demographics financial consumers perceive that anyone within a traditional FSBO is just a guilty as those named in the media.

You can see an analogous consumer reaction in how the phone companies used to operate—AT&T, MCI, the Baby-Bells. You had monolithic and very segmented people who were controlling all the markets. They were viewed as non-innovative, lacking customer forecasting and believed that if they didn't offer "it", it wasn't of value. These captains of industry dominated the telecommunication markets, but were completely out-maneuvered with the rise of mobility, and now most are controlling the back-end of the markets (their current offerings were spawned by their M&A activities).

Telecommunications was a market that for decades hid behind the regulator they blamed for their non-innovative offerings. When the industry deregulated, as the FSBO world will likely do, their customers fled for "anyone but them". Think about it, if telecommunication deregulation of a commoditized industry did not take place, would we have ubiquitous wireless solutions (i.e., wi-fi), Facebook, Google, Amazon, Alibaba, or even Apple? Where would mobile technology be—still carrying around the briefcase cell phone (and we owned several of the first Motorola cell bags)?

We traditionally believe of commoditization in terms of product. For FSBO leadership teams, they will say, "This product is not a commodity, it's rare, or custom. This product is a commodity because there's a ton of it, there's no differentiation." Yet as FSBO's, we may need to look at commoditization not in terms of what we're offering, but in how we appear to the consumer. Whether they look at us as just one of many options available to them from traditional FSBO's instead of an only option due to a unique solution capability, a layering of composite offerings into an efficient stack, or the extensive linkages to blackbox partner solutions all encased within a common security and privacy envelope.

It used to be that the community bank was the place that you went to borrow money for your home[lvi], where you went to get your child's first savings account, or where you set up your bank accounts once you had your first steady job. Our parents used to purchase CD's and other investment products from the local branch, as most did not have access or understand Wall Street. Just two decades ago, small towns in America didn't have an Edward Jones or corner store with an investment bank branch in it to make informed choices. You did your banking with your banker and the options were limited. Banks helped you with financial decisions and matters as part of a personal touch adding value and understanding to (at that time) a complicated, non-commoditized set of limited processes.

Over the last decade, with the explosion of FinTech firms, the idea of anyone firm controlling all supply chain market aspects is fantasy. Another reason for an inability to control the markets and why commoditization is already cemented within the FSBO's is a recent statistic showing that over 70% of Millennial consumers would rather go to the dentist than listen to what a banker or financial advisor should say. Across the wide diversity within FSBO channels, what could be worse than that if you are a traditional FSBO provider?

Additional studies also note from across consumer segments that most consumers firmly believe that existing FSBO's have nothing unique to offer. Also noteworthy is that over 60% of existing FSBO bases desired "something else" when it comes to solutions and products available from their existing financial institutions. Some within FSBO's look at this and say, "Why is this a big deal?" Moreover, consumers also indicated by a 3 to 1 margin that they would rather consume financial services (even lending) through a mobile app, peer-to-peer payment (P2P), web portal, or ATM rather than deal with a person during a set window of availability.

If you look back 10 or even 15 years ago, the whole idea was to create a portal by which you could keep consumers on your site. It was of secondary concern if your FSBO had value-added products and investing. So, banking leadership created their own infrastructure, hired their own engineers, sought out all resources needed to control a customer's experience and data. They were "the place", the one-stop where a financial customer could come, and in the process, get all their needs met in one institution.

Why was this? The reason is very simple. As bankers, as IT professionals, we thought our work would define our actions for two to five years ensuring a place within the organization and across the minds of steady-state consumer. When we started assembling infrastructure integration within our enterprise and our back office, we started putting all these products and these data components together creating a grand mosaic, which with time, would provide differentiation, tightly coupled systems and eventual profitability.

After all, we had time as the competition was in the same place we were. Yet today, there is a dichotomy—bankers take market risks, but are highly conservative when it comes to infrastructure (i.e., middle and back office) and control modernization. They consistently roll the dice for massive profits, but want to invest the minimum to continue to serve existing clients, as they view systems and delivery means as offering only indirect improvements to "the deal" itself.

For business leaders and IT program managers, the concentration remains on budgets, on the handoff from one app to another, and the interfaces (and their update timings) to ensure data integrity and consistency. We said, "Yeah, that was a lot of work. We need two to five years to make sure we get an ample payback." And for a while, this formula of consumer control worked.

However, consumers no longer accept the bank "reality" as an absolute—control, one-stop shop, or even favorite of industry cross-selling. Stated another way, can you imagine having your phone for five years? Consumers don't. Like AT&T and Verizon, they offer the consumers the ability to continually receive the latest and greatest components along with vastly improved security and privacy measures on those mobile phones every year—and more and more, in under a yearly commitment as part of a "refresh".

We're setting our FSBO infrastructures, our consumer relationship contracts, and our relationships from outsourcers to employees still within this timeframe—two to five years. We're trying to lock everybody in. We haven't changed our cultures and mindsets with the rest of the world. Guess what? The consumers aren't locked in. Then you look at other declining or unfavorable statistics especially across the world of financial commoditization. Why as banks or financial services do we think we're so far ahead of the delivery equation and new competition?

To shed light on these statements, the World Economic Forum (WEF) published that the United States ranks last in all developed countries in retraining its workers and getting them ready for all the new technologies and things that are transforming how society, business and personal interactions are practiced. If you look at that on a per capita GDP income basis, that's astounding. You have countries like Spain, some of the European Union, and even

China, well above the United States.

As we begin to think about regulations, governance and industry changes during the next four to eight years, how many more of these challenges are going to open up FSBO's in the United States to commoditization, disruption, and eventual failure? All the same tech players who are in the various hot spots of technology, they're going to come in and say, "Hey, we want to be part of the financial world and financially interact with consumers beyond the transactional purchase. We perform similar financial value for our transactions and those across our served industry segments, and we don't need to necessarily be a bank. We can offer services without all the burdens and regulations—and without nickeling and diming the consumer for fees. We can create value-added stacks, and let the traditional FSBO's deal with the back-end plumbing."

It is going to be an interesting time moving forward as non-traditional competition examines how they have already contributed to financial market commoditization, and how they could expand their base, while gaining broad customer loyalty. With solution pieces being compartmentalized and readily proven (across delivery capabilities in the markets and across platforms), we think it may end up accelerating what is viewed as commodities within financial services hastening the demise of many—unless the regulators send them a lifeline (with a sharp hook on its end).

A Deeper Dive: Solution Investments Do Not Equal Profits

Across investment firms and within boardrooms, the names and excitement with emerging firms has clearly overtaken those established during the last memorable expansion cycle of 1997-2001. It spawned large numbers of incubators, spinoffs from universities, and carve-outs from corporate "emerging technology" groups that spurred the NASDAQ to record highs and million-dollar VC commitments for ideas drafted by smart people.

Billions were "invested" by hopeful financial firms (e.g., Atomic Tangerine) seeking to change the competitive landscape—they didn't want to be left behind. Having personally been involved in that cycle, we can say much of it was eventually written off along with a desire to "redesign the enterprise" using emerging technology. History tends to repeat itself.

That was over a decade ago. Since then, the euphemism of "financial engineering" decimated bank risk taking, ushered in "War and Peace" style regulations and cost hundreds of thousands of those involved in the last "technology investment" cycle their careers.

Since 2012, it is noteworthy the number of financial services incubators, innovation teams, patents and technology teams that have been created within the organizational walls (collectively labeled FinTech groups herein).

Yet, in a time of declining bank valuations and a significant worry of a future recession or slowdown, bank spending on joint arrangements with FinTech firms or custom (patentable) internal developments have risen from $2.9 billion in 2013 and projected to be over $8 billion by 2018.

In fact, these projections are very conservative as other statistics already have 2017's investment exceeding $20 billion—an 18-fold increase over seven years. One can readily expect that over 90% of these FinTech-like "investments" will be written off to zero—and let us hope they were not capitalized on the balance sheet.

Already, the top FSBO's have announced they are increasing their budgets over 20% for cybersecurity, app development, blockchain (i.e., distributed ledger), and other corporate initiatives tied to digitization of the financial base. It is safe to say, we are as an industry entering the next phase of rapid technology investment—larger scale, larger budget, greater expectations and greater risks.

The Likely Impacts of All This

Consumers selecting a non-FSBO for their financial needs is like going to the prom with one person and leaving with another. You hate to say it that negative, but we think this analogy is an accurate one. FSBO's have spent many years and with great investment, courting the financial consumer to a point where they are able to accept, and even understand sophisticated delivery of products and interconnected, global markets[lvii]. FSBO's have invested a great deal in accommodating shifts in behaviors only to be rejected as out-of-touch and often with vitriol animosity. They have invested a great deal on many fronts, and it has made these brand-name players arrogant and incredulous.

In the FSBO leaders mind, they're going to try and find ways to lock them in—control the offering through incentives, through fees, through conversion limitations, through restrictions, and through breakup or early termination clauses. (Just like the U.S. phone companies lock you in contracts that other developed economies don't use.) The connected consumers are going to answer these lock-ins with their choices, and who they choose to share their financial information.

What's going to transpire is you will see a trickle turning into a flood over a period of 12 to 24 months as switches are made, new products are introduced, and margins on standard financial offerings shrink to historical lows. Consumers are already starting to say, "You know, I can go to a Simple which actually has a very good set of mobile based products, budgeting and digital management solutions" for my personal accounts. "Why do I want to pay a FSBO for the privilege of using my money and collecting ridiculous fees? What do I owe them?"

You're going to see consumers, just like we do with phones using apps and free services, being offered solutions specifically developed to easily move financial footprints from institutions A to institution B. These customer driven FSBO's and non-FSBO's will bring over the concepts that made Apple, Amazon, Google, and many others successful into the financial world empowering the customer and further commoditizing FSBO's who believe that their model of operation, lock-in, and profit honed on years of tradition will continue to work.

The real solution assets (i.e., non-commoditization), if FSBO's were leading innovators[lviii] and disruptors, would be to internalize information services within all functions (allowing for orchestration of the best mix) tailored to consistently serve a behavior transforming the financial consumer.

Financial data, some would argue, basic financial data has become commoditized and that commoditization has let others come into the market and say, "Hey we know what a general customer profile is. We know what a banking product looks like. We have all this information, and we know how to generate transactions. We can put them on the payment processing system. We know all this, how to do it, because it's become very standardized." However, FSBO's have much more and the capability to invest in truly innovative solutions, if they separate the commodity from the transformative.

When you have standardization, which FSBO industry segments (e.g., mortgage, securities) have pushed for, information sharing has made some views common and others unique based on internal or proprietary elements gleaned via algorithms or complex data streams.

A few of the past and pending examples of how information commoditization is advancing consumer agendas can be witnessed in check settlements moving from days to 24 hours to nearly instantaneous. You're seeing stock settlements that are going to go from trade plus three days to trade plus two or trade plus zero (equally plausible). A lot of the axioms that people held near and dear with the availability of information and the acceleration of technology, are beginning to commoditize what made FSBO's feel very differentiated even in a highly regulated environment. As mentioned earlier, we can also point to stock trading as one of the things that was commoditized that we thought would never happen as trade fees went from hundreds of dollars to single digits over a 10-year span.

It's that simple. No FSBO is a fortress due to the inclusion of a defensive strategy—it is a sure-fire way to promote comprehensive commoditization as technology advances and consumerism expands.

A Deeper Fear: What Will You Do?

Increasingly, as banks spend millions securing patents for their share of technology innovation, they also lock themselves into their investments, thereby limiting innovation and new ideas from sprouting in the future. Moreover, it also sets up FSBO's to be more like patent trolling firms and less like banks as time goes on, all in an effort to protect (or secure) income at the expense of recognizing and capitalizing on rapid, iterative cycle changes across consumer profiles.

Think of it this way, if we spend 18 to 36 months securing a patent or patents, how likely will senior management be to consider alternative solutions thereby making "legacy" patents a barrier against rapid market changes in currency, lending, settlements, research, and so on? Moving to understand the existing and future role of FinTech teams is a critical principle for measuring contribution across the enterprise.

- What happens when AI expands, vast data analysis becomes a compartmentalized solution and virtual reality interfaces and apps take wider adoption (i.e., Varian Rule) against the "sure bet" of smart devices?
- Are banks designed to become FinTech prosecutors against patent licensing as part or in lieu of their established charters and investor demands?
- What do we think the regulators will do with market changes, and will be it classified as anticompetitive and monopolistic after uncounted millions spent back-stopped by taxpayer

promises?

- Are conservative bankers going to become less risk averse using internal patents or solutions from "their FinTech" groups, or more if they become irrelevant for those non-banked or underbanked (representing nearly 3 billion people globally)?
- What will consumer advocacy and governance groups do when FinTech offerings contribute to discrimination practices (built into the software, apps, and layered technologies)?
- With high valve skills sets in short supply, are banks diluting their focus and returns by trying to be something they likely lack the leadership, self-destruction resolve (i.e., to destroy what exists in favor of what will grow). and a culture to not only obtain necessary personnel. but to keep them within a culture that will remain highly regulated and risk-adverse?

There has been a great deal of fear and mistrust sown over the last three years in media articles regarding "banks being left behind", "Banks becoming irrelevant against new technologies", and "FinTech firms and retailers becoming banks".

A significant amount of these themes can be found in well-written professional services articles and from interviews conducted at tradeshows and association events. But it is a case of following the boy who cried wolf, and instead becoming a lemming heading into the sea?

Can Bankers Hold onto the Consumer's Financial Data?

The most important question FSBO's can ask is, "Are we asking the right questions?" Additionally, should institutions hold onto their data thereby depriving competitors of any value from their customers (i.e., if I have it, you don't)? Or maybe the better question is what data should FSBO's hold onto to ensure they have a broker or "bookie" financial supply chain role guaranteeing them of fees and profits regardless if the consumer makes or has money? Or will a more egalitarian culture promote profitability and growth by making the institution value-added, and thus the preferred partner for existing and new FSBO collaborations? What this says for FSBO's is that depending upon strategy and capabilities, some of the approaches will be moot without substantial transition and or internal revolution.

Exclusivity for any financial consumer's data is not going to be in the FSBO's control, or an optional service put in after the development of an app, to limit sharing information with partners or even information brokers who made the offering possible. Additionally, and moving forward, there are use and delivery segments that are going to be in the consumer's control—not the FSBO, their partners or anyone but the end consumer. These consumer controlled taxonomies of data are going to be dictated by privacy and security, which the consumer is comfortable with and not how much the FSBO requires to conduct their transaction, nor for future target marketing. The shifts in how, why, where and for what purpose consumer information is being used and stored within an institution is proof that a FSBO will not be able to exclusively control the customer dialogues and relationships.

Banks always thought they were a one-stop channel and once they had you—they had you via lock-in cultures. That's how the whole idea of house holding, cross selling, and upselling approaches came to be. If we can lock them in, "we can sell them more products" came to be a theme introduced shamelessly into board rooms, and subsequently how profits were booked.

The same thing can be concluded with fee-based incomes. This strategic narrative was a key critical assumption for banks, and it has been for nearly three decades. Those aren't necessarily going to be the drivers or the solutions which contribute materially to EBIT moving forward.

If FSBO's try to put up fences to contain a financial consumer, that's going to drive existing and future customers further away. They have to be able to determine, "What markets do we want to be in based upon our brand, our solution sets, our delivery capabilities and our skill sets? If we don't have the capabilities and we don't necessarily want to be in this market then our delivery and branding models should reflect that. If we want to be competitive, then what will it take to excel, recognizing the cross-industry models are gaining extensive success with behavioral adjusting buyers?"

Often, FSBO's are chasing and spending money on things on which they were not good at or ever will be good at unless very significant investments in infrastructure, personnel, marketing, and compliance are made. That is where products and services within the banking world have garnered a lot of these FinTech interest and subsequently are going to competitively offer. These startups are going to come in and say, "We'll give you this as a banking service. We'll find various partners, we're beginning assemble these things based on best in class." Maybe because it's just popular with the consumer—that is what the consumers want, that's how they use it. It is not necessarily that it's the best, we can think of as bankers, but it is what is being demanded.

That's really the choice for bankers. Are they going to try to dig in and hold on, put up regulations, deal with lobbyists, and go to their politicians and say, "Pass legislation to keep these people out"? Well what's going to happen is if we go into that quasi-"protectionist" modes, be it regional or national, the rest of the FSBO world isn't thinking that way nor will they follow (or go quietly into their assigned corner). There's going to be vast pockets of people and their wealth that aren't going to accept protectionism, barriers to entry, or politicians who will once again be viewed as protecting the privileged within the FSBO elite.

The scary thing is that there's pockets of influential leaders who do think that way. For example, and picking on another service-based industry, there is a coalition in New York City that made up mostly of the cab companies, which are pushing for a law in New York City that will ban autonomous and self-driving cars for 50 years. Why? They ascertain that it would be too damaging to the employment status of too many people in that city. On the surface that just seems foolish. Yet, they're going to try and seek injunctions and implement laws to stop competitive pressures, which when examined in depth, are a direct result of a commodity (i.e., how to drive a car).

FSBO's could try to try to invoke similar barriers in the interest of say, data security, or public privacy, or to reign in fraud. FSBO's could try to hold onto the information that would keep them from becoming a commodity using legislation and false-positive consumer causes. It is likely those attempts are doomed to fail, while fueling even greater animosity (if that is even possible).

If you try to dig in, if you try to put up legislation, we believe anti-competitive barriers are a dual edged sword. You have FSBO's that are sitting around saying, "We don't want regulation." Well, be careful what you wish for, because that may happen. Once the laws or barriers to entry are in place, FSBO's will find that if the regulator's not enforcing compliance laws, then you've already started setting president from a legal standpoint for challenges—anticompetitive, monopolies, foreign competition and anti-capitalism. If the regulator is enforcing anti-competitive, high cost barriers, then public sentiments will turn on the political establishment, and the likely retribution for politicians and FSBO's will be very negative.

What is happening is that with a merging of digital venues and tightly coupled apps, there is a growing swell of bright and motivated individuals moving into traditional and emerging FSBO offerings. These would-be competitors go to many digital conferences to learn and influence decision makers—and they have the capital to act upon their vision. They don't understand banking but that's not necessarily a drawback.

These nascent competitors are going to come in and

say, "Well if you won't do it, maybe we'll do it—partnering with a foreign bank or domestic operator seeking to expand." That's just how the way this will work. Disillusionment in trying to deal with the process could force many to get a banking license, or more likely, find a licensed partner establishment willing to take on a commoditized industry and artificial barriers.

Commoditization in banking products brought on not so much by the products will sell, but in the way, they are sold. Commoditization is also brought to the consumers perception simply in the way FSBO's conduct themselves and use their influence to protect their historical profits.

To the completely disconnected audience, the existing barriers to entry are not a perfect inhibitor to competition when it comes to protecting markets and the consumers. These new competitors will find a way to compete with banks regardless of the inhibitors—regulators, licensing, consumer groups, politicians—as they see financial opportunity with commoditization across the FSBO's and their supply chains. What can traditional banks do to protect themselves?

A Deeper Enlightenment: A New Slope of Progression

It is not to say securing patents to protect an innovative method of operation is ill-advised, but it also has been done in a prior commoditization cycle with equal enthusiasm. However, to promote growth and innovation, FinTech's must recognize and accept their own limitations, biases and cultural boundaries when applying globally widespread digitization to their products, services, processes and iterative adaptations.

To the point, FinTech's need to be enveloped by

demonstrative boundaries.

- How will these investments in team and technology be leveraged or pared in down-cycles or irrelevant pursuits?
- Beyond quick marketing wins, how will arbitrage of solution setting be accomplished and by who?
- If there is a global or non-banking leverage for an invention or patent, who and how will this be accomplished beyond the inventors for maximum returns?
- How are investments (capitalized or expensed) to be dealt with to reassure investors and corporate sponsors that "not invented here" doesn't turn into "not our priority"?
- What filters and processes will be applied to eliminate digitization biases?
- How will FinTech's be rewarded, measured and disbanded (repurposed)?
- How does the use of external patented layers along with compartmentalized building blocks of technology, consumer experiences, joint ventures, KPO and BPO, and data isolation mesh with internal (sometimes directly competitive) standards and patents against delivery solutions?
- Who resolves FinTech issues along with command and control disagreements against business line manager demands in opposition to the "speed of smart device innovation adoption?"

There are many others, but as we said from the beginning, bankers already have a history of enveloping the hype, while discounting the eventual dissolutionment and financial consequences. Like an angry political constituency, consumers continue to confound banking strategist due to their "thinking they know their customers."

Final Thoughts

Comparable to politics, FinTech will soon experience their own revolt, commoditization and widespread write downs just as the once invincible brands like IBM, BofA, and even Facebook lose their luster—as consumers shift their attention and behaviors to something new. Moreover, the brands of FSBO's are also held in the lowest regard among industry rankings and consumers, investors, regulators and media outlets will punish these "unanticipated consequences" with extraordinary zeal.

The bravado of "we know better than anyone else" recently raised its comical satire were a CEO stated that big banks are driving the pace and path of smart banking, technological customer acquisition and rapid innovation above everyone in the industry—and even outside it. Funny thing about those leaders that survived the Great Recession, their positional retention has fed the egos and arrogance within their ranks. We have been on this threshold before.

The follies of not accepting the commoditization risks, framing the investments, and the size of market retribution, are just marking time by doing their pushups in the parking lots—waiting for the reality to take hold. Let's take our own inventory and account for realities—before this excitement turns ugly.

Commoditization is the 800-pound gorilla that most FSBO leadership teams lose sight of when they concentrate on margins, profits, and market shares. Factor in what has been transforming operations for over a decade—the comprehensive unbundling of solutions, changing of standing contract terms, progression of consumer technologies—and the methods and techniques used to combat commoditization, are beyond antiquated and unresponsive.

Finally, if banks begin to do this (i.e., dig their heels in rather than accept market demands), what's the reputational risk? How do FSBO's deal with that? We noted in the last chapter if you have 30% to 40% of the people willing to switch their bank to a Google or an Amazon immediately if they could offer that then, why wouldn't you?

Illustrative Challenges / Opportunities

Category	Illustrative Challenges / Opportunities	Controls and Implications (The 4-V's) Volume/ Rate	Variety /Adapt	Velocity/ Freq.	Veracity/ Accuracy
1	Commoditization requires reformation of both structural and fiscal compositions beyond one-off improvements.	↑	↑	↑	↑
1	Need to adopt short-sequence innovation life-cycles (i.e., experimentation) to keep enterprises from falling behind.	↑	↑	↑	↑
1	Exponential rise of linked data and financial information on discrete customers providing vast opportunities to create new value-added offerings beyond traditional FSBO's.	↑	↑	↑	↑
1	Outsourcing has become about the management of commoditized offerings as transformational / iterative innovation targeting customer growth becomes competitive advantages.	o	↑	o	↑

Category	Illustrative Challenges / Opportunities	Controls and Implications (The 4-V's)			
		Volume/ Rate	Variety /Adapt	Velocity/ Freq.	Veracity/ Accuracy
2	With FSBO offerings sounding too similar, customers concentrate on price rather than value thereby reducing loyalty and brand to limited leverage.	↑	↑	↑	↑
2	Commoditization is aided by strategies focused on domination or being the largest, rather than targeted approaches, which yield the strongest offerings.	o	↑	o	↑
2	Traditional reliance focused on the specific purpose app—which has been commoditized—resulting in limited distinction contributing to the rise of FinTech.	o	↑	o	↑
3	A failure to understand customer value has reduced the barriers to entry.	↑	↑	↑	↑
3	An inability to anticipate and predict future offerings in an anti-establishment political environment.	o	↑	o	↑

Category	Illustrative Challenges / Opportunities	Controls and Implications (The 4-V's) Volume/ Rate	Variety /Adapt	Velocity/ Freq.	Veracity/ Accuracy
3	Focus on risk-aversion / attribution has reinforced a conservative FSBO culture as reputational risks surrounding innovation, privacy, security and knowing customer needs become secondary.	O	↑	O	↑

↑ – Increasing non-linearly, unpredictable behaviors
O– Increasing but predictable

NOTES

"We succeeded in turning around the business ...
just in the wrong direction."

Image ©Shutterstock, Inc.

5

Regulatory Non-Compliance

"We rely upon self-declaration, upon what is presented to us by a bank's internal management. But often they don't know what's going on, because banks today are so vast and hugely complex. I don't think I have ever been deliberately lied to – though obviously, I might not know about it. The real threat is not a bank's management hiding things from us: it's the management not knowing themselves what the risks are, either because nobody realizes it or because some people are keeping it from their bosses." – **The Guardian**, Voices of Financial Services blog.

The topics covered in this chapter will likely see the most changes through 2021. With a proposed reinstatement of a "new-age" Glass-Steagall Act, to a repeal of numerous provisions within Dodd-Frank, to defanging the CFPB (or even declaring it unconstitutional), to simply just not enforcing what many within FSBO's deem government overreach, the likelihood that regulatory compliance costs[lix] and efforts will decrease has helped push the S&P banking index up over 30%. There are widespread corporate expectations that most of the post-Great Recession measures will be removed. It seems like FSBO's will have much to celebrate in during the remainder of the decade—assuming the chaotic world of cross-market risks don't somehow change the unproven, euphoric math.

Contrary to post-recessionary consumer demands, the largest five U.S, banks now control over 45% of global assets (a fourfold increase in eight years), FinTech firms are applying for banking charters, and consumers have even more methods to connect, pay for, and manage their lives than in recorded history—all at their mobile fingertips. Much has happened to move the needles of what banking and finance must deliver as part of federal, state and local regulatory compliance in just seven short years[lx].

There has been, as Napoleon once said, "... a foot on fire[lxi] ..." driving stoic organizational cultures to accept new realities that run counter to ingrained traditional beliefs due to new and pervasive, or invasive regulation, depending upon your FSBO position. Compliance regulation has indeed been a double-edged sword in curtailing profitability, but also improving risk and soundness of survival.

So as the regulations (or shackles) are set to be adjusted or removed, we need to prepare for the flip side of their amputation. If the direct or indirect drivers of change are removed, will the innovative data stimulus also wane in favor of traditional approaches more comfortable for bankers? And finally, without a sense of urgency and competitive confrontations across and within financial channels and consumer demographics, will traditional bank thinking once again become status-quo merely augmented by creeping innovation?

A Deeper Reality: Responsible (Regulatory) Innovation

With the turn of the millennia, the idea of responsible innovation (RI) surfaced primarily in the European Union (EU) as part of their Horizon 2020 program, and within the United States among academics. In general, RI seeks to balance responsibility, values, authority, sustainability, ethical, policy, legal, social, accountability and corporate governance from cradle to grave for innovations (for a primer, see **Responsible Innovation: A Primer for Policymakers**, Brookings Institution).

Blended together from hundreds of pages of text, it places a "burden" on the inventor or originator of an innovation (or innovative solution set) to determine the boundaries of its use, and to help with the enforcement across its deployment life. (Think of how the inventor of gun powder might feel centuries after its positive and destructive use.)

More recently, this iteration or transformation of how innovation need to be responsible has been raised by the OCC (Office of the Comptroller of the Currency) in its recent paper, "**Supporting Responsible Innovation in the Federal Banking System: An OCC Perspective**". Whereas RI is now an emerging discipline, it begs a fundamental question—who decides what is "responsible"? Is responsibility the same across all demographic segments? Is what is being responsible today, irresponsible tomorrow? Who is going to determine adherence to RI, with what criteria, and penalties against stepwise or evolutionary transformation?

While it seems evident that the OCC has embarked on RI to determine and lessen risk taking by institutions and those who in proxy serve their agendas. It is, from what is written, about ensuring proper supervision and stability across technology advancements, which are likely foreign to stoic government agencies who do not want to be blindsided as they were back in 2006-2009.

RI provides a promise of a framework to regulate FSBO investments and deployments. As Thomas Curry, Comptroller of the Currency delivered on March 31, 2016 at Harvard Kennedy School of New Directions in Regulation stated, "*Not every innovation is appropriated for a regulated financial institution, and not every innovation that is appropriate for a regulated institution is appropriate for all regulated institutions.*"

Mr. Curry continues, "*A responsible innovation is one that is consistent with sound risk management practices. That means the bank understands the product and the risks it carries, and has the capacity to manage those risks.*" He concludes, "*And finally, responsible means that it complies with laws and regulation, particularly those aimed at protecting consumers.*"

When these comments are merged with the actual report

outlining the eight principles for OCC action, it is the last three principles where the "devil is in the details"—encourage banks to integrate RI into their planning, promote dialogue with formal outreach, and collaborate (on RI) with other regulators.

So, there is a question for FSBO leadership, "Is RI going to help when placed under the filters of a regulatory agency who is seeking to redefine their scope of influence?"

Recognizing the Challenges and Risks of Repeals

There are just over 6,000 U.S. banks under the watchful eyes of numerous regulators and consumer watchdogs. The media headlines portray FSBO's as out-of-touch institutions seeking to roll back the controls put into place following tens of billions in bailouts that kept them in the lifestyle they were accustomed. As noted, the top five domestic institutions by asset size dominate nearly 50% of retail markets leaving over 5,995 institutions to fight for niches.

Therefore, with the projections that tens of billions in banking capital could be freed up and put to work across the economy, the question of who will benefit from this needs to be addressed[lxii]. For example, will small banks have the capabilities and balance sheet to stimulate loan growth—or will it be just a few institutions who reap the windfalls? Our current thinking about regulatory easement and a growing debate on non-compliance going forward is that these billions and billions in economic windfall may not be as true as hoped.

When you look those institutions completely dominating the markets, against a simmering debate of too big to fail (TBTF[lxiii]), the political and regulatory debates start to move into moral accountability. Taking off the shackles of regulatory compliance, scaling back conflict of interest rules, redefining capital and risk retention rules, and allowing greater lending sounds like something FSBO's would greatly benefit from—if the spree is not followed by delinquencies and defaults. These loans (via unencumbered capital) could be used to stimulate private investments in infrastructure (which is currently targeted at $800 billion over the next decade).

If the financial journals are right, and FSBO's stand to gain over $100 billion with rule relaxation, it isn't likely going to be the small or even regional banks that increase balance sheets and profits. It's going to be the larger institutions and the disclosure of who and why they received this windfall will likely have negative impacts on FSBO brands and political careers. TARP was hugely controversial along with the (continual) soap operas surrounding the GSE's, and it will not be surprising if when risks rise, and the increased leverage turns sours, that those same political cheerleaders begin to carry the banner of tighter regulations and TBTF[lxiv].

Will the results of easement or non-compliance be reminiscent of the downfall history between 2003-2007 which laid the foundational bricks for the Great Recession? We firmly believe that part of these regulatory retractions (i.e., repealing the "worst" of Dodd-Frank) will quickly pass (a relative term for political institutions who cannot typically make up their minds) through the U.S. House of Representatives. However, when it comes to the U.S. Senate, we look for a slower debate and vote (especially when consumer protection and capital controls start to weigh on debate and amendments). These areas aren't necessarily going to affect those smaller FSBO institutions, the under $5 billion ones, and we think that's something which is going require watching as we move from hyped promises to mundane reality.

So often within our FSBO industry everybody talks compliance. Applications are out there for compliance and when Dodd-Frank was framed and deployed, new and established vendors brought out their tightly integrated data solutions, automation and validation at the source, and exception processing tied to rules-based risks engines (to mention just a few).

How much of that is going to be valid moving forward? FSBO's spent billions getting ready for the last eight years of regulations (and SOX[lxv] in the decade prior). The results beyond "safeness and soundness" has made FSBO's accurately access information, assign risks, reduce speculation, and turn the institution's operating procedures into a competitive asset that wasn't even contemplated before regulatory burdens. Are FSBO's going treat these advancements as a sunk cost, which are written down, or a stepping stone to emerging business models that moves beyond dogmatic fees and minimally valued products and services?

"Well, if we get rid of all the compliance mandated layered dashboards, prediction metrics and standardized measurements that have quantumly changed the intelligence covering our assets," then you must ask, "Are we basically shooting ourselves in the buttocks just because we don't want to do it?"

Yes, regulatory compliance is complex, burdensome and expensive to implement. However, we also lose all the intelligence of what the market's doing, even down to the individual loan that may be wrapped up into some type of securitization instrument or vehicle. The risk with the blind removal of regulations, assuming there is political capital available for repeal, is that what will FSBO's put in the place for the insights lost and the future litigation, which will be initiated with the next economic down cycle? After all, the U.S. public won't be as forgiving to anyone if some official or FSBO says "bailout".

A Deeper Dive: It's More Dispersed than Centralized

When looking at the effects regulations can have or organizations, we note that there is a minimum of 16 areas directly and indirectly impacted by their existence or removal. Several of these impact areas have strong ties to disintermediation trends—payments, market provisioning, lending, funding, and contact—all which are wrapped around critically important envelopes of security, privacy, risks and disclosure. As noted in the illustrative figure below, each impact area can be ranked (before and after regulatory issuance) to establish markers for what happens when or if the expected wave of repeals is issued by Congress or executive orders (including the forthcoming implementation or delay of the Volcker Rule[lxvi]).

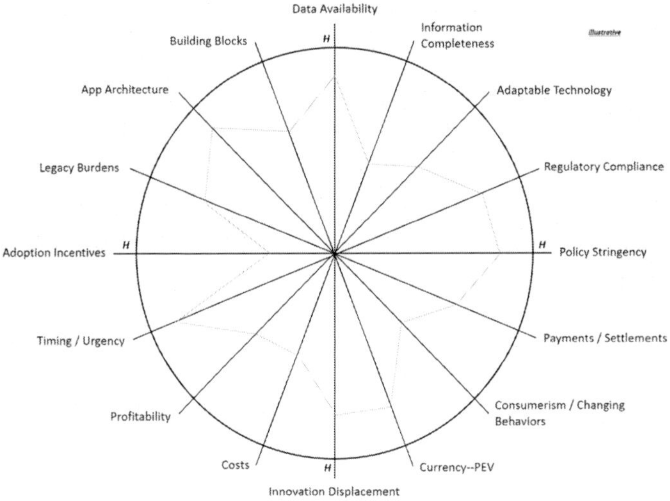

But regulations of the last decade have been huge and complex, and the lost momentum, risk-mitigation and cross-purpose improvements that were to be gained from their streamlining, might also deduct from the upside of their removal.

As we know, the financial services supply chain is multifaceted, layered, compartmentalized, and globally connected—make one change in an area and we cannot reliably predict what the up or downstream impacts might be within a dynamically expanding environment (i.e., across supply chains).

We should not forget how we got to the state we are in today—commoditization, animosity and financial supply chains. While it seems refreshing to celebrate a reduction in regulation, good for the corporate soul, it might be a lost opportunity to transform and further disintermediate what we already have arrived at. The "baby and bathwater" analogy comes to mind.

Identity the (Ugly) Beasts and Deliver Common Themes

This market controlling regulatory beast, a seven-headed hydra for this discussion—one head Dodd-Frank, the second CFPB, the third stress-tests and living wills, the fourth Basel III, the fifth attributable risk and cybersecurity[lxvii], the sixth domestic oversight and governance and the seventh everything else—has increased compliance costs between 300% to 700% for every FSBO budget since the start of Great Recession just a decade ago.

Layer on internal and external auditing diligence and legal disclosures (and adherence, reporting, and data) and the costs rise even further not to mention a dearth of talent and skills needed for updating and maintenance[lxviii]. From a common-sense regulatory approach, "we" settled on "boiling the oceans"—not providing the isolation or risk compartmentalization that drove the initial intents for a "safe and sound" financial system.

Indeed, from most common-sense approaches, regulatory moderation is something that needs to happen sooner than later to reduce burden, while promoting comprehension, eliminating duplication, and enabling global competitiveness.

Still, for bankers and a governing administration that believes lending is curtailed for, "...so many people, friends of mine, that have nice businesses and they can't borrow money" may not reflect the reality of market actions[lxix]. For in this instance, as noted in Bloomberg, these statements may test reality especially when commercial lending has risen by over $2 trillion USD in just four years—double what it increased the eight prior years. So, what else do we factually know about the actions within our regulated environment against what is stated or perceived?

As we know, interest rates have been held historically low by the U.S. Federal Reserve (Fed), and are still below adjusted measures before the housing meltdown. As we know, housing supply[lxx] is also at a multi-decade low (estimated to be approximately four months). As we know, homeownership is currently bouncing around 63% (the lowest per the U.S. Census since 1965) and off the heights of over 69% back in period between 2004 to 2006.

And as we know from various financial sources (e.g., Bloomberg), the Fed may start to unwind (i.e., stopping reinvestments) the $1.75 to $2.0 trillion in MBS's with a downstream consequence of raising interest rates, while reducing GSE purchases—and perhaps hastening the latter's repurposing. Finally, we know from outlet like the WSJ that with farming income at $63 billion, half of what it was just four years ago, a likely outcome may be a substantial increase of land availability for household formations and homebuilding. Even farmers, of which one of your authors is, needs to eat and pay bills (this group represents the next, biggest set of bankruptcies forthcoming with some economists predicting these defaults will total in the tens-of-billions due to extensive leverage and complex unwinding).

It seems, that if regulations are going to be relaxed for the stimulation of homeownership[lxxi] and supported by responsible borrowing, then conditions must be set to provide a resurgence of private market and capital opportunities where risks are recognized and priced accordingly.

After all, with the Baby Boomers declining and millennials sidelined by an anemic recovery over the last decade, there is an opportunity for the 78 million Millennials to take a larger economic responsibility including home loans, which are currently being written at a rate of two to one for women over men. What is seldom debated, post-recessionary regulations, is that the transparency of what FSBO are producing is far greater than just five years ago. Therefore, the elimination any regulatory compliance burden will be weighed against financial product transparency, safe and soundness—even before attributes of consumer protection and derisking are absorbed.

So, this is where the FSBO's find themselves. At the beginning of an anticipated, prolonged easement cycle that may be timed with a generational demand for rapid products (i.e., private securitizations) driven by pressures that may push originations well past the $1.5 trillion per year (not including refinancing).

Broken down further, if refinancing holds at the 10-year norm of 40% to 50% of all loans generated—as compared to the period from Q4 2008 to Q2 2013 where refinancing was 50% to 70% of the total (see Realtytrac)—a prolonged lending cycle under a regulatory easement may significantly enlarge opportunities for lenders seeking an expanded base of clientele. Superficially, it seems nothing could be wrong, and that regulatory easement may have found "legs" in the face of need[lxxii].

A Deeper Reality: Compliance is Designed into the Business

It took years, decades to craft the rule sets today—many of which were designed for a finance world that doesn't exist domestically or globally when it comes to consumer behaviors, money flows, data availability, research and reporting, and well, nearly the entire financial supply chain (e.g., virtual and digital currencies, cloud computing, 24/7 market trading, KYC solutions, credit risk management and mitigation, and many, many more).

It essence, by the time a rule is written, it is usually for a commoditized market offering—not for innovation and consumer advances. One could make the argument that regulations as conceived today is folly when applied to the technology and data capabilities in existence.

To solidify the previous statements and trends, regulatory easement (and of course, compliance), should be viewed as part of a mosaic of factors as shown in the diagram following. When mapped out against these illustrative characteristics, a coverage area can be discernible for impact, deficiencies and investment. This approach, when applied with objective assessments, can demonstrate before and after projections thereby properly gauging impacts against existing operations and proposed initiatives.

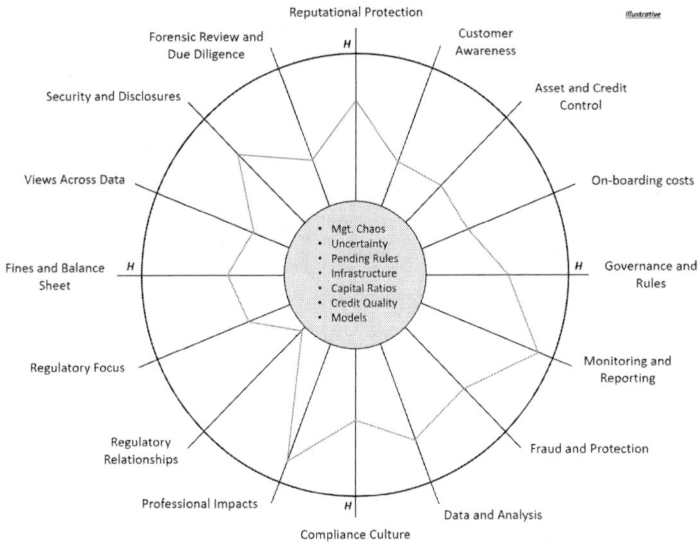

On top of everything we have touched upon, or perhaps because of, it is estimated that over 90% of FSBO's should modify their business models—due to rapid changes in the world and political landscapes.

Less Regulation is Not a Panacea for Less Complexity

Less regulation is not a panacea and more is not derisking—neither opposite pole approach (i.e., spread across a continuum) is very realistic and usually never a sustainable growth area for capitalistic economies. These "poles" are great for political theatre, but fall short when it comes down to implementation. For example, we have heard rumors that if the GSE's were privatized and removed from government conservatorship, the U.S. Federal Reserve (Fed) would sell their stakes freeing taxpayers from the burden and indirectly expanding the federal deficit by trillions of held assets propping up a sluggish, post recessionary economy. In this example, in MBS's alone the Fed holds between $1.75 and $2.0 trillion USD (as of this publication) in limited-performing securities—roughly one-half of its current balance sheet and four times what it held before the Great Recession. We wonder, if the Fed hadn't indirectly added these to a $20 trillion USD government debt[lxxiii], what would have happened to the world's largest economy when domestic and global financial markets as well as the U.S. Congress froze?

Let's put this in the perspective with some hard numbers and an example. Home loan originations, true originations not refis, are still right around a trillion dollars, that's it (at their peak years before the Great Recession, they ranged from 250% to 430% greater than today's run rate). The rest of the market continues to run at over 50% refi's because everybody's still trying to lock in the rates guessing where the interest rates will be once the Fed winds down their portfolio and (hopefully) repurposes the GSE's.

If the Fed starts winding their portfolio down, and the only way the Fed's going to be able to wind that down or repurpose GSE's, is you're going to have to offer something that doesn't have an implicit or explicit guarantee (thus lowering borrowing costs)[lxxiv]. Mortgage lending without guarantees will demand many basis points above T-Bills. The problem is, if you go to private capital, you're talking hundreds of points above T-Bills resulting in a need for higher rates and the elimination of spreads between government and private issued or managed securities (for bundles of whole loans).

Because nobody's guaranteeing it. Where's the stop-loss? Who's going to take first loss on the mortgage securitization (assuming we're not including covered bonds and other on-balance sheet items). If they go back to traditional non, qualified MBS of any kind, and if U.S. government T-bills remain at 3% or 4%, then what rates and divisions will a private or public loan holder be guaranteeing within these asset ring-fenced portfolios? Well guess what happens? You start unwinding that, the market's going to demand interest rates that are well beyond the traditional 15 or 30-year instrument—substantially more.

All the topics we talked in the prior chapter, if the Fed unwinds their positions (and which they need to), will place a heavy burden on paper liquidity and rates—they cannot continue to double their holdings of mortgage-backed securities buying them from GSE's or FSBO's to mop up excess supply. There must be a reason to keep the market somewhat liquid and reinvesting in semi or illiquid assets?

If the housing market continues to generate only $1 trillion in true originations, there's a lot of downstream consequences if we're simply moving government guaranteed assets from one government entity to another. We don't think a lot of people have thought through the phenomenon of why this continues to occur years after the Great Recession was declared over.

On the other hand, what do you think the chances are that we're going to see that move towards privatization? As most of us have read in the headlines, we know there are bills pending and common threads being continually discussed within Congress and the U.S. Treasury. One of these ideas, just introduced, will take away the government's right to use GSE income for government spending, which basically renders this a lot less valuable to the federal government to balance budgets or fund other initiatives. If it is made into law, it would force the ending of conservatorship and release some form of the GSE functions back into the private markets.

Do we think the chances are high to see a move toward privatization in the term either via the GSE repurposing or with a coordinated push to restart sluggish markets? We don't think you'll see it until probably 18 to 24 months after a governing administration takes office and only if there is a mandate to achieve. We firmly believe you're going to see a lot of people talking about it (as they have since 2010). It seems everyone has an opinion on it now, but as the government nears a decade of control over the GSE's, the debate polarizes on ideological grounds rather than financial value and long-term liability.

Some people will like it—some won't—but the changes must be debated and a viable and methodical series of iterative plans must be proposed. It's going to be tested. It's going to be challenged. It's going to be modified, and we can tell you that the consumer advocacy groups and right to home ownership groups, they're going be part of these repurposing initiatives to ensure that NO hint of explicitly or implicitly discriminating against individuals or families. This GSE example may become the litmus test on why less regulation and oversight is not a panacea for simple functioning markets and a reduction in risks all within a single FSBO category—mortgage securitizations (but a key component of redefining the FSBO supply chains, see Chapter 9).

A Deeper Displacement: The Impacts of Compliance

The reality is that regulatory compliance has triggered widespread innovation displacement (i.e., traditional advancements forced out by cross-industry behaviors). It also harbors future disintermediation events that are not even being contemplated as the cycle of disintermediation moves from a perceived one-off event to that of continuous disruption brought forth by cross-market urgency, new uses of data, and multiple adaptable technologies.

- Removal of Urgency: Banking and its bankers have been their own worst enemy when it came to defend themselves against overreach. It was a problem rooted within corporate culture and internal politics—something that has been changing due to mandated regulations, stress testing, living wills and a host of federal, state and local guidance. Yet, there has been positive mergers and acquisitions (the highest in a decade), profitability has been restored along with consumer confidence, and the pace of technology change is robust. What will the plans for 2017-2020 look like when the impetus for change is gone? Will politics and dogma once again force a future cataclysm?

- Reduction of Data Forces: The timing couldn't have been worse. Just as financial services organizations and operations perished under bad loans, exotic instruments and a host of assets not properly risk attributed brought down many enterprises, the exponential explosion in the volume of data overwhelmed the remaining crippled operations. With the flood of data sources—social, retail, financial, professional, personal, and mobile to name but a few—the potential to create maps of behaviors, forecasts of consumption, bottom-line profitability when properly integrated provided insights, bespoke products and services deemed before the crisis as a technological dream. Yet, data integration brought forth by the demands of complex regulations has provided a foundation for new FSBO architectures and offerings that will likely force into retirement traditional front, middle and back office systems. So, if the removal of regulations usher in a decline of complex, multi-faceted data analytics and taxonomies (e.g., machine learning), what will happen to the expertise assembled? Will the innovations being made be slowed in favor of nostalgic BAU?
- Failure to Apply Adaptable Technology: The most popular topic of the press and within enterprises involves blockchain and its cousin distributed ledger. Yet, there are also advanced technologies that have expanded greatly in the last five years including artificial intelligence/predictability, cloud provisioning, "robot" advisory solutions, machine learning, cryptocurrencies, and biometric security to meet regulations, while streamlining complexity

and costs. Moreover, additional adaptive technologies include intelligent apps, IoT linkages, virtualization of solutions and platforms, mobile applications, comprehensive digitization and service based architectures and many more. Many of these advancements have come at a time of burdensome regulation. While we are not making a proclamation that these advancements came directly because of regulation, they were likely advanced to their adaptable states by the need to satisfy complex and in some cases, cross-purpose oversight. So, what will transpire when the burden is gone? With the strides made subside or disappear as investment wanes and fragmentation resurfaces?

Regulation is indeed a double-edge sword. In a highly-regulated industry such as financial services, it is very tempting to jettison the new especially if it comes with an association of heavy regulatory burdens. It seems many institutions are already booking results and readying rollbacks even before the initiatives have run their course—it might be a mistake. Unwinding far-reaching regulatory demands will indeed result in a short-term bump regarding profits, discretionary budgets, litigation and available hours.

Regulatory Easement Requires a Laser Focus

It seems from politicians, regulators and media reports that repealing or easing of proposed or existing compliance has become the expertise of anyone willing to take up a banner against overreach. Others shout, "No, No, No! Look at all the people in “this class” that were disadvantaged." Then the reply becomes, “Yes, they're in that class. Are they being disadvantaged because they're in that class or are they being disadvantaged because the market bears what the market bears.” And the discussion then turns to acrimony and animosity usually with FSBO’s in the crosshairs as they are the nasty boogeyman of what's wrong with financial markets.

Janet Yellen, Chairwoman of the U.S. Federal Reserve on August 25, 2017, at the annual Jackson Hole financial symposium opened her speech (all 19 pages worth), with the following:

> *"A decade has passed since the beginnings of a global financial crisis that resulted in the most severe financial panic and largest contraction in economic activity in the United States since the Great Depression. Already, for some, memories of this experience may be fading--memories of just how costly the financial crisis was ....*
>
> *A resilient financial system*[lxxv] *is critical to a dynamic global economy--the subject of this conference. A well-functioning financial system facilitates productive investment and new business formation and helps new and existing businesses weather the ups and downs of the business cycle. Prudent borrowing enables households to improve their standard of living by purchasing a home, investing in education, or starting a business. Because of the reforms that strengthened our financial system, and with support from monetary and other policies, credit is available on good terms, and lending has advanced broadly in line with economic activity in recent years, contributing to today's strong economy.*
>
> *At the same time, reforms have boosted the resilience of the financial system. Banks are safer. The risk of runs owing to maturity transformation is reduced. Efforts to enhance the resolvability of systemic firms have promoted market discipline and reduced the*

problem of too-big-to-fail. And a system is in place to more effectively monitor and address risks that arise outside the regulatory perimeter..."[lxxvi]

What she indicates at the start of her speech and continues, is exactly what one might expect from the most powerful Central Banker in the world, be careful and methodical before a knee jerk reaction is undertaken to remove regulatory guidance. It took the better part of eight years to craft and implement the latest round of regulations—and much prosperity hinges on a balanced approach especially if something goes wrong.

However, we think that's where sensible controls start to leave their rails, and it becomes part of the reason the regulations were "done the way they were to punish these SOB's next time something like this happens." The problem with this punishment mindset is that the "next time" may never happen because it's going to be a different market consuming very tailored financial instruments. It's going to be something else that goes wrong as products, technology, and globally interconnected instruments are created—so will those rules necessarily apply? We thought we had the "correct" rules before, and most were surprised and angered when the void of ignorance caught up to the reality of increasingly tightly coupled financial supply chains.

It's going to take some time to get this through, but at the same time, we think there's going to be a lot of incremental changes—good ones. Call it people trying to do things for the right reasons. Some used to classify these good deeds as (old school) Letters of Intent, "we want to do this, but how are we going to make it happen?" That's where legislative processes come to bear and partisanship takes a backseat to global realities of how financial markets perform in a world that never shuts down, and is connected by satellites and fiber optics to nearly every corner on the planet.

And yes, there is a rule-making process,

implementation mechanisms, and limits to consider. There are countless committees and special interests, which need to be heard from and their concerns addressed. Ideas could sit there for years (e.g., a U.S. covered bond market which never took hold), or it could find roots and be out onto the floor for a vote in months. And yes, we have not even divided into the reality of state and local rules, and what might be waiting in their legislative processes or seeking to challenge federal guidelines in the court systems.

Even so, it's a ticking clock and FSBO's need to be aware that regulations are not easing to the extremes that they likely desire. We believe there are three areas that need to be a focus for FSBO's in the coming months and years.

First and foremost, who is your customer and what type of customer is necessary to achieve the primary business model—first and foremost. You can't be all things to all people, nor do you want to be. The other thing FSBO's must realize is not everyone can only service the affluent customer or even a general classification demographics like Millennials. Everybody wants some classification, but the reality is how are you going to carve out your spot based on your business model and capabilities?

Secondly, whatever that business model is, whatever segment you think you're going after, how are you preparing to differentiate yourself especially when factoring in regulatory compliance demands? As an example, are you going after fifteen-year loans? Are you going after a twenty-year loan, which markets believe will become the new standard, what are you going to do? Are you going to get out of car loans or get into them? What are you going to do and how does that go?

Ask yourself, "Can I get enough technology wrapped around that (insert your FSBO target here) to make it simple or reach the desired base? Can I get enough data, encapsulated with or by technology, that will it be able to support my intelligence, my dashboard, my forecasting accurately? How fast am I going with this? Am I going to basically flying blind into the wall?" You've got to get that information together and basic things like loan to equity ratios to determine what the chances are, what's this consumer doing, based upon all this financial products and potentials. Are we seeing the consumer starting to pay late? How are we getting all that information? Are they over-extending themselves and putting their fragmented information across the various databases? The whole concept of big data with learning technology is going to be one of the key points, as it underpins not only the business model, but the need to meet or exceed regulatory guidance.

The third point FSBO's need to do, given those first two directions, given where we want to be, given how we should make that happen, what do we need to do in terms of personnel (and subsequently as an institutional culture)?

Additionally, as we review America versus the outsourcing market, where are these outsourcers going to get the talent from in the face of visa debates and onshore talent utilization? Well, if you don't have it already within the borders, there's only two choices—import it or (re) educate the personnel you have (or want to hire). If we can't find the people and can't make the cost business case, based upon the first two points, this third point is going to be the deal breaker. You may understand what you need to do, but how are we going to make that happen? AI and machine learning can take you so far, but then the question becomes who manages and enhances the artificial intelligence when you lack the capabilities internally? What will it cost? How will we support its introduction, its adaptation and what always is "surprising" the cost of ownership?

A Deeper Analysis: Asking What Easement Will Impact

Our belief is that because very few individuals were held directly responsible for the demise of stoic firms and individual families, the rules written since 2009 were about extracting "a pound of flesh". However, these retribution principles morphed into something else becoming themselves vehicles for channeling of anger against those deemed elite or "untouchable".

The easement of regulations brings an unintended whale tail risk that could have huge economic and political risks—which is why easement and not direct repeal will be the likely norm. And even with fast intent, the delayering of complex processes, technology, people and data will take months and months as most institutions already have made huge investments.
In general, the promise of regulatory easement begs a few additional questions when reviewed against the above.

- What will easement do to risks, compliance activities and personnel mixes?
- If easement or rollbacks are achieved, how will the organizational transformation initiated by this hydra be impacted?
- What will downstream consequences become if consumer facing rules are rewritten (e.g., if lending is spurred, what does this do for balance sheets, private securities, new asset instrument classes)?
- How will these reductions of regulatory compliance impact sizes of institutions (e.g., community banks versus systemically critical)?
- Will the insights having been derived from complex data collection lead to less operational improvements, actionable intelligence, and risk attribution? That is, if we remove what we had been directed to complete, are we blinding ourselves?

We are experiencing a great deal of uncertainty across our society and within our various industry sub-segments. We still have bad actors doing nasty actions to unsuspecting consumers and investors. As a collective industry, we have sought to drive out asymmetrical information spurred on by a heavy hand of the regulators. Our operations are now some of the most advanced in the world due to a need to automate for cost efficiencies—many spurred by regulatory compliance.

Yes, regulatory easement is needed. Yes, we need to apply common sense to our financial supply chains. And yes, easement provides a huge windfall of avoiding in solutions that simply comply with a regulation.

But no, easement cannot represent the elite of Davos regenerated. Easement brings an opportunity for FSBO's to quit looking down, or in some cases, over our shoulder. We cannot drive forward fixated on our rearview mirror. We cannot pretend the old days have once again come and party like the privileged—no matter what that Tweet said.

The Changing Insulating Realities

With recent business articles projecting that FSBO's will receive a windfall of up to $100 billion after the rollback of Dodd-Frank, our brains spawned an image of a banker, clean shaven, blue suit, white shirt, tired face, a Davos ticket stub in his suit pocket, and a bold, red necktie with a champagne bottle in one hand and a lending contract in the other dancing to Prince's lyrics of, "... we're going to party like it's 1999 ...".

After FSBO's spent tens of billions fighting, settling and influencing multi-agency legislation, the new U.S. administration has signaled its intent for a multi-year effort to "tame" what is portrayed as numerous unfriendly, government overreaches. Simpler would be nice, but billions have already been spent in preparation and compliance.

This calls into focus a fundamental perception—can we really achieve any meaningful regulatory easement or repeal? Is the market going to bear some form of non-compliance? Moreover, will the chief risk officers (CRO) tasked with identifying and mitigating risk supported by the compliance officers came to meet the I's and the T's of the rules, become secondary to the deal makers seeking to carve out new territories? Doesn't that really demand a change to the business model, especially when you start factoring in how people use that technology, rules, and disclosures that add value and define investor confidence?

When you come to understand and interpret that

we're going to need to meet the regulations and the extensive changes across our consumer bases, we frequently don't have the people (or the organizational cultures) necessary to achieve the desired results. It's a complex process to find or even partner with qualified talent, much to the surprise of many, as organizational hiring has become a factory mindset that has consistently failed. It can be attributed to the changes needed to define, manage and run an FSBO, which has moved from simpler linear decision trees to a forest of options and requirements that are tightly coupled with regulatory compliance.

Those are impacts that are going to be more significant now that we've got kind of people trying to tighten up immigration and put limits on special visas. It could be very challenging for FSBO's to find the people they need across the regulatory compliance spectrum.

Today, somewhere between 25% and 35% of the total domestic financial chain is staffed or satisfied by foreign personnel. The exact percentage dedicated towards regulatory compliance varies widely depending upon the business segments the FSBO serves, but these individuals consistently work at 70% or less of standard rates of an onshore domestic person, and often much less when "lifted and shifted" offshore. That's a big difference in costs spread out over a quarter or a year. If you start attacking the differential and put in "America First" it creates another set of regulatory burdens and justification of "can we compete?" So, while the desire is to reduce regulation, the introduction of workforce restrictions can place new demands even as other ones are removed.

The regulatory realities, especially when it comes to innovation, has once again taken center stage when it comes to reinventing all aspects of FSBO's. Regulatory innovation, which was once dominated by automation of processes (i.e., front, middle, and back offices) for efficiency savings, has permanently changed. Process design and reengineering made popular by historical experts such as Davenport, Drucker, Peters, Hammer and Champy has evolved in a response to regulations and significant changes in consumer behaviors because of advancing technological capabilities.

Additionally, regulatory realities with cognitive advancements or machine intelligence coupled with ubiquitous, compartmentalized apps layered across the financial supply chains, are reaching a tipping point once thought fantasy—promoted by visionary individuals such as Raymond Kurzweil (noted author, inventor, and Google advisor).

As FSBO's, regulatory compliance has turned us into technology firms with financial offerings, destroying what is commonplace, and dealing with uncertainty across cultures (and regulators) that thrive on certainty. Is it really that surprising that central bankers, economists, and flip-flopping politicians are hard pressed to find solutions that last more than a few months?

However, what is the outcome of these "advancements"[lxxvii]? What are the paths forward when mapped against existing infrastructures and practices? Can a cohesive, less reactionary plan of compliance be created which promotes growth, while avoiding further marginalization of unbanked or underbanked? Does the industry really need more regulations—to protect business margins and models from non-traditional entrants—while they secure patents, align with partners and source innovative solutions?

If we focus on non-compliance as our industry banner, sentiments across segments will likely turn negative and we once again will be seeking shelter in the arms of the very groups that seek punishment. It's time to focus on the future.

And yes, we ask a lot of questions.

Illustrative Challenges / Opportunities

		Controls and Implications (The 4-V's)			
Category	**Illustrative Challenges / Opportunities**	**Volume/ Rate**	**Variety /Adapt**	**Velocity/ Freq.**	**Veracity/ Accuracy**
1	Those who are indecisive surrounding current and future regulations will limit their capabilities to leverage existing investments, while alienating consumers and markets looking for leadership	↑	↑	↑	↑
1	Data availability has expanded transparency, and mitigated risks while discretely identifying vulnerable asset classes, and produced thousands of opportunities to create bespoke solutions to address a "me" consumer market.	↑	↑	↑	↑
1	Compartmentalized data solutions coming from non-traditional sources have spawned an industry for information exchange that can be used for market, financial, and competitive intelligence.	↑	↑	↑	↑

Cate-gory	Illustrative Challenges / Opportunities	Controls and Implications (The 4-V's)			
		Volume/ Rate	Variety /Adapt	Velocity/ Freq.	Veracity/ Accuracy
1	Adaptable, machine driven solutions have provided a series of solution platforms that will usher in utility or BaaS delivery models that will favor those eager to shed legacy solutions, while providing accountability and auditability.	o	↑	o	↑
2	Incremental and revolutionary consumer consumption patterns will permanently transform front, middle, and back-office models all originating from regulatory compliance demands—often originating from outside the FSBO enterprise.	↑	↑	↑	↑
2	Innovative displacement will supplant the newness and relevance across traditional innovation deployments resulting in a non-steady state of IT(O) and BP(O) deployments.	o	↑	o	↑

Cate-gory	Illustrative Challenges / Opportunities	Controls and Implications (The 4-V's) Volume/ Rate	Variety /Adapt	Velocity/ Freq.	Veracity/ Accuracy
2	Legacy burdens will continue to consume over 80% of traditional IT budgets until enterprises can create core capabilities around app (and application) rationalization, archiving, transformation and retirement.	O	↑	O	↑
3	Creation of compartments or building blocks of solutions internally and across partner delivery networks will provide clear advantages in quality, delivery, market acceptance and cost containment.	↑	↑	↑	↑
3	An ability to sustain urgency across complex and divisive business opportunities provides incentives for iterative consumer and regulator satisfaction.	O	↑	O	↑

↑ – Increasing non-linearly, unpredictable behaviors
O– Increasing but predictable

"Let's shrink Big Data into Small Data ...
and hope it magically becomes Great Data."

6 Digitization

"If you owe the bank $10,000 you have a problem. If you owe the bank $10,000,000 they have a problem. After all is said and done, more is said than done. Government does not solve problems; it subsidizes them." – **Ronald Regan**

Digitization is, if looked at holistically, a primary reason for compliance challenges, a key factor in determining future credit risks, forecasting consumer behaviors, and a bullseye on the challenges with restarting private securitizations (and transforming the GSE's). Yet, in our quest to chase the Holy Grails of "All Things Digital", the "best and most profitable customers", and "operations which exceed the expectations of regulators and investors", have we embraced digitization solution sets simply because it is digital (e.g., big data, social media)?

As we stated previously, tomorrow's digitally linked world "isn't your father's computer science, it's not even close". Digitization is nothing new (heck it is representative of computer science cornerstones dating back nearly 40 years ago)—its acceleration and pervasiveness is where adaption, integration, organizational, cultural, and implication challenges make FSBO winners and losers. Digitization is fundamentally about what you do with it—explicitly RESULTS achieved!

Furthermore, how information is gathered, processed, merged, extrapolated, retired, transacted, protected, managed, strategized or anything in between cradle-to-grave regarding the sourcing chains of all financial industry's products and services, is likely inadequate when assessed against transformative performance standards. For the last four years, chasing the goals of digitization (and realizing measured and consistent returns) has been comparable to chasing a free-floating apparition.

In a corporate era defined by aftermaths and results against consumer behaviors and sovereign economies, where is digitization heading? How will it impact our current operations? What are the pathways to embrace for assurance of success when contrasted against where we are today? How will our FSBO be able to layer, in complex and dynamic models, elemental digitization (which is where we are today) into solutions that span internal enterprise capabilities and into the realms of collaborative digital orchestrations?

And one more thing, the principle of cradle-to-grave information management is nearly meaningless against the morphing of data, let alone the 20-fold increase expected in the next six years. How can an organization "keep up?" Inside sovereign borders that restrict electronic access (e.g., China, Russia), where does the tipping point of financial safe-and-soundness meet free-market and "moral" restrictions to promote growth—or is it at the edge of a cliff as data becomes more important than those seeking to strictly control it (or their ability to do so)?

Yet, regardless of society, individual or blue-sky prejudices, the objectives necessary to guide the integration of digital solution sets into corporate infrastructures (including cloud deployments) surround four main categories—efficiency, profitability, consumerism and insight (i.e., intelligence). These markers need to be explicitly defined for every enterprise seeking a digital vision and mission—be it be in mortgages, capital markets, consumer lending, or even digital media. It should be noted, that this chapter is supported by three years of research, and we've included the originally released synopsis is **Appendix A**.

Copious Green Fields

For technologists and developers of software, the last decade has been thrilling as the science of data and its generation has spiked new corporate disciplines including machine learning, big or vast data, advanced AI, data discovery, business analytics, market repositories (assembled for identification and resale), machine-to-machine interactions, Internet of Things (IoT), sensor science, mobility advancements and efficiencies and the list goes on. Holistically, global IT spending has reached a near $4 trillion annual spend (i.e., telecommunications, hardware, software, services and data) growing at a normalized, conservative rate of 3% to 4%.

While some professionals label these advancements as (positive or negative) disruptions, risks, outcomes, structural shifts, consumerisms, innovation, or as convergence in delivery, whatever it is called the quantitative high-water marks continue to exceed year after year projections.

Somewhat out-of-phase from the initial breakthroughs, (re) birthed out of necessity and innovative emergence, come cybersecurity (i.e., white hats, black hats and crime), multi-level (learned) compliance, consumer privacy, data obliteration, digital sources and uses, value measurements (and loss of benchmarking), brand impacts, intellectual rights management, governance and government rules of engagement, rich-media management, storage management, integration and quality assessment software, cyber legality, digital transformation and a hundred others in advanced stages that weren't even out of the R&D labs (or basements) just two decades ago. Effects of these can be seen in a few statistics:

- There are annually up to 1 million cybersecurity positions that go unfilled (source, Internet Systems Security Association—ISSA), and consistently rising.
- 75% of firms believe they need emerging (loosely defined) data analysis skills, while nearly 45% cite

time and resource deficiencies as critical impediments (source, SAP).

- By 2020, B2B[lxxviii] and B2C internet transactions will exceed 450 billion per day (source, IDC).
- Along with the meteoric rise of analytically educated data FTE's, expenditures on IT services will soon break the $1 trillion mark (source, Gartner).

Beyond digital statistical one-offs, CSC (now DXC Technology) discusses a collaboration consequence labeled "cocreation". In it, companies come together to create digital ecosystems within "insurance, payment systems, healthcare, transportation and more" collectively sharing the product and service benefits. These "disruptive" compilations open new opportunities for markets and profit, while transforming employee skills, time to delivery and of course, customer expectations and usage.

This disbelief of events by some in FSBO's is precisely why in a post-recovery world many FinTech startups have found homes with non-banking institutions seeking to break the cast models surrounding lending, securitization, insurance, credit scoring, closing[lxxix], servicing, and of course, capital markets. Digitization for FSBO's, like presidential politics, is less and less about dynasties or birthrights. Digitization is the future of FSBO's, and if the culture is ready, it is a critical factor for an enterprise's deployment, success, and profit

A Deeper Dive: It's Really About a Culture

Lessons learned when applied against FSBO's whose public reputation is still very low, highlight six critical areas where culture can impact the ethical integration of digitization solution sets.

1. Whereas procedural changes will be needed to integrate internal, joint venture, partner and third party digital technologies, staff and even tightly coupled outsourcers must have a working

knowledge of what is being changed, why, and its impact to day-to-day and corporate strategy.

2. Landmines of discrimination, lending practices, product promotion, rates, credit worthiness and even insurance and closing costs can be created when the need for urgency outdistances the ability to understand the implications of digital solutions.
3. Accept that the digitization of everything (DOE) is a journey, and not a readily implementable end-state. Cultures honed on periodic large initiative sprints may lack the flexibility, rewards, motivations, political dynamics, and even review processes necessary for a patchwork of programs some outside the direct control of internal leaders.
4. Not everything can be accomplished "tomorrow" even if it is a great idea—conservative cultures just cannot adjust. Moreover, the upstream and downstream impacts of regulated enterprises cannot be underestimated, but the incorporation of market facing and operational efficiency digital solutions cannot be ignored. An aggressive balance needs to be postulated within budgets, reward processes, and between peers to ensure that the best possible outcome is achieved in the shortest timeframes.
5. Accept that just because a competitor is offering a digitized solution does not mean you should clone a response just because it was that way in the past. Digitization has a "nasty habit" of creating innovation even among the most conservative or well-run institution (and even their regulators). Factor in consumer behaviors, and technology advancements across interconnected machine interfaces and we now understand why even the Fed is looking at updating "what is" secure payments across all its venues.
6. Within a regulated FSBO, enterprises may need to move into grey areas of offerings—BUT regulatory involvement should be sought out even in the absence of specific laws in advance of release. While this may seem counterintuitive or risky, it may be needed to address non-traditional

competitors, and it is likely the best way to show good-faith and an impetus to change across overly restrictive regulatory compliance laws. For FS cultures and its regulators, it is no longer about the compliance fight—but the global market's growth and the ethical stimulation of consumer demand.

For leaders within and across FSBO's, many are assessing and planning for change using successful digitization pilots, while also conducting readiness and performance evaluations against recognized leaders such as Google, Amazon, Apple and others. Stated differently, the old is the old, the new is something else especially when it comes to layered, orchestrated digitization and the need for permanent organizational cultural shifts.

More than Just the Breakthroughs

For the remainder of the decade, FSBO's should expect M&A's driven by evolutionary digitization opportunities. In fact, it is estimated that nearly $1.5 to $2.5 trillion over the next five years will be raised for M&A events, many of which are driven not singularly by products or services, but by connecting with customers, greater social outreach and access to information for prediction.

The temptation to expand digitization to "know more", "do more", "be more" will increase as consumer digitization grows. But with continued industry consolidation on the horizons, the rationale for action needs to be measured against the key indicators that are tied to mission, strategy, goals, and objectives.

What firms and their leaders need to realize, like investors in the stock markets, is that to chase the market during a growth frenzy is fraught with hidden risks and downside demise. Chasing the digitization revolution requires:

- A clear and comprehensive link to strategic intent and directions.

- A focused purpose so that upside and downside boundaries can be established including assumptions and limitations.
- A defined list of what taxonomies (and categories) the digitization serves so that trends can be properly framed, reconciled and even benchmarked.
- Personnel and processes fully integrated across current and future channels (i.e., channels, products, services, and support).
- Acceptance of the boundaries and retirement of digital artifacts (including regulatory demands).
- A pipeline of what is next and when, who, and how they will be released.

Just because it is digital, doesn't make it accurate. Just because it is extracted and manipulated across machine learning algorithms doesn't mean that the result will produce success—lasting or iterative.

What is concerning, especially when people across FSBO mention, bubbles, over-extensions, risks, and a host of other technological challenges, is that chasing digitization without purpose and clarity is analogous to the market crashes of 2000 (technology) and 2007 (housing).

"We're dancing" as a prior CEO said, just before he lost his job, but are we confident in why and how? Are we chasing it for competitive reasons? Are we chasing digitization because of media reports? Are we chasing it because our employees and focus groups believe it is the future?

All the reasons are valid—if we know why, if we know how, as long and we know by what means the pieces efficiently fit.

A Deeper Strategy: Making Digitization Worth the Transformation Pains

Taken in its entirety, what is coming into focus for FSBO's is that those who have the underlying data and digitalization strategies will prosper in the transition from automation to digitization (i.e., the second and third waves of business transformation respectively).

Challenges, in their simplest form, focus on prediction, intelligence, protection, growth, indicators, profitability and most importantly digitization relevance.

Yet, many valid questions surface across the executive suites:

- "What strategies are likely to ensure success? Are these initiatives tied to performance, and if so, can they be quantitatively linked?"
- "Where should investments, partnerships or acquisitions be made to transform current operating methods?"
- "How do we tap into existing or emerging consumer behaviors to shorten time-to-market, while promoting brand, product, and service adoption (and adaptation)?"
- "How can measurements serve a purpose which is emerging, based on prediction and be reconciled?"
- "Is everything being presented as digitization achieving significant paybacks? Will it? How do we avoid the 'field of dreams' traps that started out so positively?"
- "Do we have the correct personnel, skills and processes necessary to leverage hidden assets or new digital sources?"

The questions above are the common sense every management team should be asking and challenging the axioms that all which is digital has a pot of gold at its completion. They ask the questions because that is what they need to do—not because they "don't get it". They must answer to investors, consumers, public, regulators, politicians, activists, peers, and the list goes on.
FSBO's are sensitive to risks and untested approaches even though consumers are expanding their use of digital apps, business solutions, and of course, service-delivered solutions (e.g., Uber).

Try as FSBO's Might, Pandora's Box Has Been Opened

Digital Yin and Yang—the darkness and the light. Some describe this as striking a balance of duality between complimentary forces. Similarly, in a world now encompassed by enormous data output and precision decision making, there are emerging taxonomies or disciplines which are balanced by privacy versus profitability, marketing versus identification, protection versus accessibility, accountability versus irresponsibility, and fraud versus ethics (i.e., digital integrity).

When these discrete continuums (along with others) are assembled, they conjure a "picture" of how particular digitally driven enterprises operate utilizing a cultural profile to achieve results and repeatability. Simple enough, no? No. With the speed of digitization, nothing is as it appears and its complexity (along with hidden implications) are anything but straightforward especially for organizations unaccustomed to rapid change. The light-speed of digitization is everything but BAU.

Yet, it is the assembly of this puzzle of these complimentary and competing attributes that make digital adoption (and subsequent adaptation) so difficult especially within highly regulated industries such as the collective enterprises comprising FSBO's. Underpinning this evolving digital picture are the realities of organizational capability, acceptance and readiness—aka its culture and complexity of interworking.

As inferred by the failure rate and ephemeral adoption of innovative digital solutions, per Harvard University nearly 9 out of 10 technologies selected for enterprise initiatives fail and per Stanford University over 90% of the startup firms fold, the ability to cope with widespread and growing digitization impacts are materially influenced by personnel and the organizational practices for risks and rewards.

As our economies and financial supply chains

undergoing a very broad digital awakening, it is interesting to see front page headlines touting, "Increase capital and shelve Dodd-Frank". This news is generated from yesteryear leaders who forgot that if we put the Great Recession back 20 years, the lack of computing and complex digitization may have let the likes of Countrywide and Bear Stearns survive.

Yet, with the information discovery and global financial services system integrations embryonically available at the start of the last decade, losses, risk and unwinding could not be hid (perhaps if they could, buying once leading brands time to recover) as information was readily available. Wish upon wish, we can only go forward and assuming or making a case for restoration of controls forgets the methods, techniques and depth of information available to today's modern investor, lender and borrower.

Fast forward a decade (yes, we are nearly a decade from the start of the last economic crash), how will rolling back regulatory principles at a time of unprecedented access serve a benefit when it will take another decade to do so? It is like fighting for life boats when there are none. Why not recognize the impacts of digital availability and its multifold increases, and architect a system (along with governance and compliance) that meets a future—rather than assume like the Luddites that the IT world has not already begun to shattered Moore's law? And, as we all have been educated, past performance is NO guarantee of future success, so why try and think we can restore a time when technology was less pervasive?

A Deeper Review: Why Complex, Multi-Layer Digitization is Not BAU

Digitization requires a difference in thinking, different skills, and a different attitude. It also requires, in some cases, a destruction (or disruption) of the past. Not everything digital is innovatively relevant to FSBO's. However, as we have seen market and technology events are clearly set to disrupt nearly all aspects of financial procedures, products, and services. But unlike the past where destinations were defined at the end of a monolithic solution, tomorrow's journey has very little to do with either.

Digitization is a journey filled with vast growth, unstructured solutions and a fluidity that will chafe many established FSBO's. For institutions like TD[lxxx] who are investing in MORE brick and mortar locations, you can safely bet that these future branches will be digitally provisioned with nearly every high-tech solution. They will be designed to embrace customers expecting more than a simple human interface. It may superficially seem the same—but it is a radical departure.

Digitization requires skeptical thinking, but in the end, it will create a chasm between those who leverage it embracing its uncertainty and frequent transitions.

And what is more psychology than technology, just because it is Tweeted, posted as a status, pinned on a site, or anything else, does not make it real or even valuable. Digitization brings a whole new meaning to "be careful what you accept as fact."

Embracing the Digitally Unintended Consequences

Some authors have termed an enterprise's ability to adapt to digitization as "cyber cultures". With increased frequency, the days where enterprise respect was driven by office size, organizational structure, and even profitability are shrinking. If we look within our own corporations, we notice that those with access to and protection of information are commanding greater respect without all the trappings of traditional employees who "climbed the ladder".

An outcome of DOE is that it has made a shamble of traditional organizational designs and brought to the surface that those who command information, or its creation, are far more in demand than even the managers who still seek command and control. It's this chaos of design and politics that threatens inaction even when the best efficiencies are made available.

Reuters has published, citing Brand Finance as its source, that nearly $71 trillion dollars (i.e., half) of corporate valuation is comprised of intangible assets across 58,000 firms[lxxxi]. What this rapidly growing number eludes to is that corporate cultures will have to be more about information and digitization rather than traditional benchmarks, asset management or even historical communications and investor reporting. While some economists will cringe, consumption and economic prowess is about information and how it is captured, managed, manipulated, archived, and even destroyed.

A Deeper Dive: Management Practice Solution Framework

To combat this unintended consequence of cultural chaos introduced by digitization, proactive and leadership organizations have begun to embrace change using emerging management practices. A few of these include:

- Anticipate errors in the roadmaps, while reacting and adjusting without repudiation. Mistakes will be made and as a result when using untested, new, or even unfamiliar solutions (i.e., experimentation)—within enterprises that spend years using the same software banking vendors. Iteration is good, short delivery positive, but discrete multi-year maps may not provide the value they once achieved.
- Employee hiring, satisfaction and evaluation is moving from the mechanism of slotting candidates to ones of behavior and flexibility—and above all a passion for change and life-long learning.
- Popular management consulting leaders for decades advocated, "If you can't measure it, you can't improve it." True, yet as we have seen with Amazon and before that GE, creating data and

following it using unadapting algorithms can lead to cultures where non-conformance is rooted out and disposed. Yet, innovation, relevant innovation along with embracing the unknown (i.e., non-conformance), is what has made firms very successful. In a rapidly digitized world, what we think we know from past actions is what causes the most growth stagnation.

- Old engagement and cross-division boundaries are disappearing, while personnel integration with systems are rising (i.e., a seamless blending with the very systems they utilize). Just as Bloomberg ran a 30K word article, "**The world belongs to people who code. Those who don't understand it will be left behind**[lxxxii]", the implications of a digitized culture will pride itself on delivery, information accuracy, software expertise, big data, and so on against a lack of structure, office space, and dare we say it, a distrust of traditional authority.

Within FSBO's, these unintended consequences will be hard to internalize. Why? As FSBO balance sheets shrink (by choice or by external actions), there is limited desire to veer from the norm and challenge the procedures and hierarchies, which some credit for a return to stability (and reinforced by regulation). In many cases the unintended consequences of digitization are too much to accept—A Queen of Hearts scream of "Off with their Heads" reverberates down the "Mahogany halls".

It's Been a Long Time Coming

Like people, sometimes organizations do the right things for the wrong reasons. In an e-commerce world of B2B, B2C, C2C, and B2G interlaced with pervasive Wi-Fi coverage, the march of structured and unstructured digital data grows at over 100% compounded per year projected to reach 60 trillion gigabytes by 2020.

To adapt and utilize unchecked digitization solution

sets (where digitized data is moving from 22% usable to 37%), pundits are increasingly recommending cross-industry operating models for existing FSBO's—incorporate innovation incubators, create software firm cultures, utilize even utilize an Amazon-like approach to serving customers (i.e., retail). It might be concluded from the recommendations that to be a viable FSBO, you need to operate as anything but a FSBO.

Empirical analysis shows they are correct—but not for the reasons provided, let alone the preparation necessary to create enduring foundational changes.

Moreover, some believe that there is no downside to digitization—just upside, and in most cases, they would be correct. So, are we getting wiser? Are we better able to predict with certainty outcomes for emerging solution sets? Is technology and its cultural integration becoming more transparent and predictable? Are we able to leap tall buildings with a single bound using layered digital solutions? (Just kidding on the last one?)

To promote digitization success, a repeatable yet expandable framework needs to be adopted to aid with the development and adherence to digital roadmaps across a high-regulated industry such as FSBO. This framework must be able to deal with "drill-down" requirements so that custom crafted solutions aligned with culture, capabilities, infrastructure and more can be achieved.

Below is an adaptable framework that provides the foundation for action plans. Starting at the top, there are five functional divisions. These divisions can be broken into three segments—knowledge taxonomies, banking functions, and digital categories. While the concentration has centered on FSBO capabilities, other industry models have been created for this approach thus delivering repeatability and cross-industry adoption.

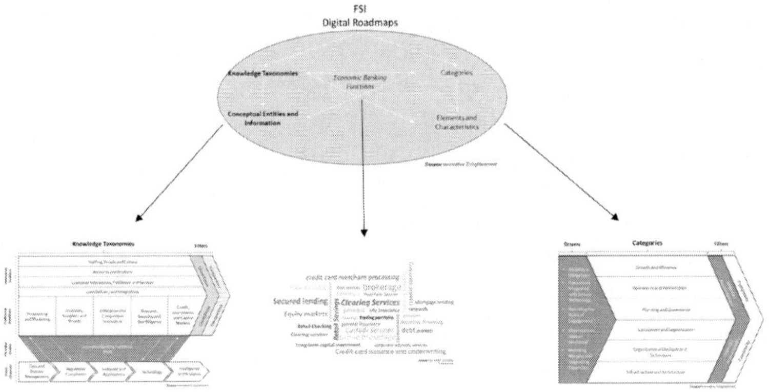

The model above and the thousands of underlying analytics represents nearly three years of work, research, and vetting into market influences and intellectual capital surrounding the very broad category of "digitization".

To prove that these ideals could possess efficacy, and of course, how to let organizations profit and adopt positively to the inevitability the models, needed discrete components, and further breakdowns. What resulted over a year and correlated quantitative values, are over 7,000 intersection points spread across 14 taxonomies (each with 14-26 sub-categories) and six grouping of digital impact segments (decomposed into 24 sub-impact zones).

Since early 2013, we've been trying to ascertain what is the secret sauce for digitization success and longevity? For nearly three years, we've assembled hundreds of artifacts and thousands of pages of published research, case studies, and recommendations spanning academics, consultants, vendors, and thought leaders.

However, as we attempted to assemble a common go-forward rationale, we realized that few ideas fit together and little of the solutions spanned more than a narrow range of applicability, or were just a macro blanket encompassing digitization inevitability.

Additionally, we believe and contrary to current tenets, that digitization may plant the seeds for the next great "depressional" shift as cross-domain workforces, consumers, investors, corporations, data and of course, technology permanently transform trade, purchasing, supply chains, and an ability to repay (or participate) in any economic interchanges.

Digitization, the very thing that makes processes faster, data increasingly accurate and pointed, and consumers clamoring for more, could provide the accessibility for foreign wars[lxxxiii], cyber terrorism and unintended consequences that cannot be solved with central banker's "extraordinary powers". With a decade nearly gone since the start of the Great Recession and ensuing financial meltdowns, could we fathom how fast economic collapse would come with the hybrid cloud solutions, data gathering, system interconnectedness, and consumer power available with just mobile devices?

A Deeper Analysis: Into the Empirical Models

For FSBO's seeking to transform their operations and enterprises, the realization of concrete criteria and factors for digitization initiatives ensures that focus and prioritization for digital customers who are increasingly demanding advanced and interconnected solutions. The diagram below is a macro rollup of the nearly 7,000 points of intersection involving digitization and its adaptability.

Correlation Legend

- High
- Moderate to High
- Moderate
- Low to Moderate
- Low / Not Applicable

Knowledge Taxonomies	Growth and Efficiency	Operations and Performance	Planning and Governance	Consumers and Segmentation	Org. Methods and Techniques	Infrastructure and Architecture
	Digital Impact Categories					
Staffing, People and Culture	Moderate to High	High	High	High	High	Moderate
Accounts and Products	Moderate to High	High	Moderate	High	Moderate to High	Moderate
Customer Interactions, Fulfilment and Services	High	Moderate to High	Moderate	High	Moderate	Moderate to High
Core Delivery and Integrations	Moderate to High	High	Moderate to High	Moderate	Moderate to High	High
Prospecting and Marketing	High	Moderate to High	Moderate	High	High	Low to Moderate
Investors, Suppliers and Boards	Moderate to High	Moderate	High	Moderate to High	Moderate to High	Low to Moderate
Enterprise and Competitive Innovation	Moderate to High	High	Moderate to High	Moderate	Moderate to High	High
Research, Sourcing and Due Diligence	Moderate	Moderate	Moderate to High	Moderate to High	Moderate	Moderate to High
Credit, Instruments and Capital Markets	High	Moderate to High	High	Low to Moderate	High	Moderate
Data and Process Management	High	Moderate to High	High	High	Moderate to High	High
Regulatory Compliance	Moderate	Moderate to High	High	Moderate	Moderate to High	High
Software Applications	High	Moderate to High	Moderate to High	High	High	High
Technology	High	High	High	High	High	High
Intelligence and Analytics	Moderate to High	High	Moderate to High	Moderate	High	Moderate to High

Source: Innovative Enlightenment

(**Note**: Each row is further decomposed 15-26 times spread across 24 vertical columns with each receiving a rating)

As we recorded these empirically into cross-linked databases, they generated a repeatable set of processes (and outcomes) that can be custom tailored to an organization across any industry arriving at a “digitization index”.

This repeatable model is not only advantageous for current and future states, but for surveys and custom generated IC that could be conducted with a “mobile first” approach and integrated with big data sources. Hence, it is a model that adapts to time and changing circumstances, while moving with firms rather than simply at the bequest of their high-priced advisors.

With the development of the models and criteria it reinforces that digitization is a journey—not a discrete destination or project to be completed and then checked off. If you think holistically, it is not difficult to see how this can be proceduralized—software, extensions, services, and marketing.

As a further example of the granularity that organizations need to adhere to as digitization maturity embeds itself into operational cultures at unprecedented paces, the decomposition of the data taxonomy into further efficacy criteria shows impacts of scale, speed, prioritizations, and even the rules of engagement.

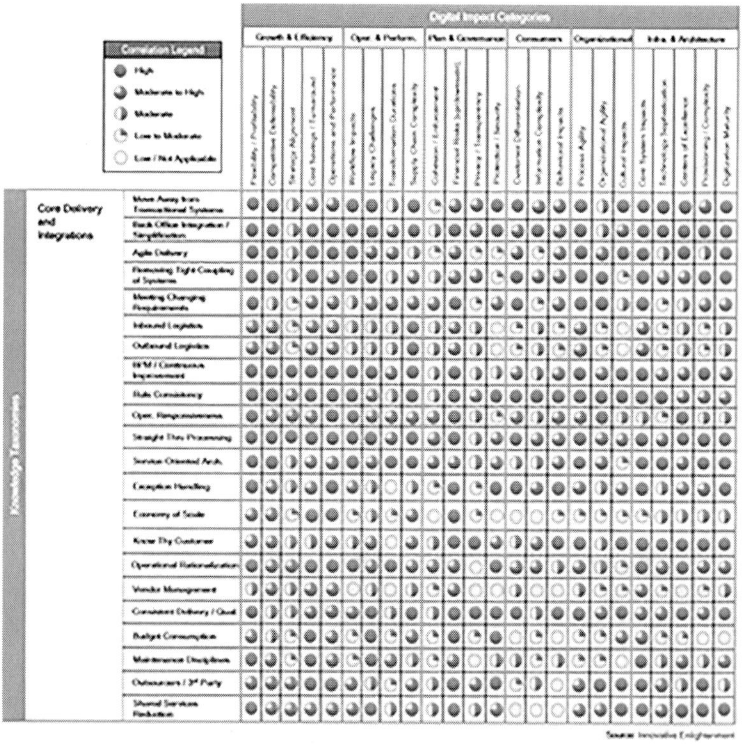

And stating the obvious, each of the sub-taxonomies within that data (and the thirteen additional segments) can be broken down further as needed to properly align with the enterprises goals and timeframes.

What this multi-year effort has shown is that digitization is a moving target that is often and simplistically just associated with technology advancements. Digitization is not, as marketers sometimes ascertain, a lock-in by demographic segments (e.g., Millennials). So, as we move rapidly forward in FSBO's with digitization initiatives that are no-longer just turnkey projects, they require especially over the remainder of the decade a realization that FSBO's and bankers are "swimming in the shallow end of a very enormous pool".

New "Leaders", New "Bands"

In conclusion, "living" digitization efforts demand the use of business orchestration to ensure that the thousands of requirements are properly fit into the cultures and operations of our existing enterprises. For FSBO's, the challenge is teetering on who is the best person for the efforts—when the reality is that it is a cohesive group or leadership team that must come together to balance growth, change, and risk.

Many will try to expand current committees, program offices, or outsource it entirely to obtain a "check in-the-box". The problem with the digitization of today and tomorrow is that it is cemented in our past—FSBO's are not set up with rewards and responsibilities for the collaboration (and trade-offs) inherent with business orchestration applied to complex, layered digitization demands. Without accepting the tenants of this chapter, FSBO's will find that their much-anticipated programs have become part of a technology trap.

Yet, if we want to survive the industry onslaught, digitization beyond one-offs should become the norm. If we want to move beyond regulations (as we know today), we must leverage technology via interoperable stacks. If we want to provide journey maps for corporate onboardings, communications, orchestrated methods and techniques (e.g., based on Agile), we must replace the hodge-podge of solutions from vendors, outsourcers, and our own staffs.

Think of it this way, much like the debate over driverless cars are being thrust into the front pages (where traditional automakers caught unprepared are casting the most doubt on the upstarts of driverless cars), FSBO's can learn from the mistakes and perceptions playing out in another industry. Driverless cars indeed are about using "parts" from many disciplines to arrive at a new solution (i.e., orchestration). And it is highly probable, that the inventors will not be the innovators—as the latter have found a way to relevantly apply innovations to arrive at a "new normal" whereas the former often have a myopic vision of the invention's potential[lxxxiv].

For FSBO's, our historical ideals are getting grey—it is time for "new leaders of the band". We could be another auto industry fighting a consumer change and an orchestrated revolution—is that the position we wish to take?

Illustrative Challenges / Opportunities

		Controls and Implications (The 4-V's)			
Cate-gory	**Illustrative Challenges / Opportunities**	**Volume/ Rate**	**Variety /Adapt**	**Velocity/ Freq.**	**Veracity/ Accuracy**
1	Developing capabilities to move from simple digitization to complex layering of digitization involving the Internet of Things (IoT).	↑	↑	↑	↑

Category	Illustrative Challenges / Opportunities	Controls and Implications (The 4-V's)			
		Volume/ Rate	Variety /Adapt	Velocity/ Freq.	Veracity/ Accuracy
1	Deploying skills and capabilities needed for rapid, iterative digitization across internal and third-party solutions.	↑	↑	↑	↑
1	Developing and incorporating compartmentalized solutions across the "4-V's", which move digitization into the more advanced, digitalization.	↑	↑	↑	↑
1	Honing and internalizing digitalization strategies that move digital add-ons into layered and isolated digital building blocks, which can be used for digital processes and seamless data sharing.	o	↑	↑	↑
2	Moving measurement and monitoring of digitization capabilities to accurately reflect direct and indirect contributions to margins and profit across adaptive solution sets.	o	↑	↑	↑

Category	Illustrative Challenges / Opportunities	Controls and Implications (The 4-V's) Volume/ Rate	Variety /Adapt	Velocity/ Freq.	Veracity/ Accuracy
2	Becoming innovatively relevant by eliminating false positives of regulatory compliance to innovation actions with the incorporation of cradle-to-grave designs encompassing less than 12-month durations.	↑	o	↑	↑
2	Designing for unstructured, vague demands across units and within IT deployments to achieve market distinction, delivery flexibility and capital outlays.	o	↑	↑	↑
3	Eliminating the impulses (and initiatives) that are chasing themes rather than addresses specific weaknesses all in the name of being competitive.	↑	↑	o	o

		Controls and Implications (The 4-V's)			
Cate-gory	**Illustrative Challenges / Opportunities**	**Volume/ Rate**	**Variety /Adapt**	**Velocity/ Freq.**	**Veracity/ Accuracy**
3	Design of programs and cultures that accept and prepare for continual, rapid product, service, technical and organizational disruption(s) on an enterprise scale—driven by consumer behaviors and omni channels.	↑	↑	O	O
3	Develop methods and capabilities across the organization and within partners (including outsourcing arrangements) that exceed customer perceptions for security, privacy and value-add.	↑	O	↑	O

↑ – Increasing non-linearly, unpredictable behaviors
O– Increasing but predictable

"What we've done is make it dramatically easier to navigate the corporate hierarchy."

7

Staffing, Training and Organization

"There is no employing class, no working class, no farming class. You may pigeonhole a man or woman as a farmer or a worker or a professional man or an employer or even a banker. But the son of the farmer will be a doctor or a worker or even a banker, and his daughter a teacher. The son of a worker will be an employer - or maybe president." – **Herbert Hoover**.

Financial Services and Banking Organizations (FSBO's) will be unable to compete effectively in markets dominated by virtualization, social purpose, and rapid innovation—they are architected and regulated for a defunct consumer representation.

Across a range of nationalistic sourcing overtones, what is striking are the vague and sporadic discussions regarding the skills, staffing, and organizations that will be required to restore a beleaguered middle class. For those on the frontlines within banks and lending operations, the competencies and strategies required to leverage the information and behaviors of an emerging class of borrowers is continually being redefined in the face of regulations, disclosures, and complex risks.

Additionally, to accomplish a fundamental economic shift of a net importer to an exporter requires an answering of numerous independent and layered questions—specifically around personnel. For example, how will employers be called upon to assist with wide segments of clustered populations or nationalities? What types of education and skills will be needed by geography? Which institutions are best suited to aid with student or worker advancement? Where will be changes impact our organizational structures? And more importantly, who will pay for this rebirth at a time when educational debt is at a historic high?

Breathe Deep, Prepare for a New Path Forward

When we look at the "needs" for redesigning the organization, we find three critical lanes of transformational requirements—staffing, training and organization. If we can take those one at a time, staffing, training and then organization, what do they have to be thinking about to fix the problems? Let's start with the first one, staffing. How are we going to address that historically different?

We briefly discussed this in several of the earlier chapters. How do potential applicants feel about joining FSBO's and what do really think about their ability to innovate, provide career direction, to provide work that's interesting? We've also mentioned that FSBO's ranked near the bottom of industry profiles, and the result, we think if you look at it, is really what we call a disconnect in their business model.

For instance, if you look at segments of the staff, they are for the most part individuals that they want to work within a financial institution. However, with changing product and service mixes, FSBO's are now aggressively seeking cybersecurity personnel[lxxxv], they're hunting for those who can create and launch successful mobile apps (e.g., ATM to mobile phones), they're hunting for people that can do large data integration across heterogeneous platforms including those resident within the public and private cloud repositories—to mention just a few. All those wonderful discrete technological pieces that FSBO's often interview for, and want to hire for, those kinds of people that have that capability don't necessary want to go into a traditional banking environment, and that is often a result of the business models and how they drive profits.

So, if you look at the systems FSBO's have historically dealt with and those which need to be "mobilized" with non-FSBO partners (e.g., checking account, credit cards, loans), the changes have gone historically well. However, they were rudimentary and a prior representation of something they were in control over and knew from experience.

Nevertheless, how do you maintain those things when they grow exponentially complex, require compartmentalization, and tightly-coupled integration across systems that in FSBO measures are in their infancy (i.e., < a year old)? What do FSBO's need to do to offer different capabilities? And most people that have the ability and the skill sets to work on those types of applications, don't necessary see FSBO's as innovative, while also not perceiving them as collaborative within their departments, across departments, and especially across industries.

For FSBO's it has always been dog-eat-dog, only the strong rise to the top of their organizations and internal politics is an asset worth more than capabilities. And so, the people that have the social skills were raised on social media, collaboration, and peer-to-peer interactions, don't often see FSBO's and their culture as providing what they value for their embryonic careers. Therefore, there's a huge disconnect on how you attract those people as compared to the way standardized employment testing, resume automation, and skills test have done over the last decade.

And the final challenge is how to do this without completely upsetting the brand and operational cultures that have dogmatically made them successful. FSBO's have not really figured out how to create a balance let alone eliminate the negative perceptions caused by their bad actor actions laid bare by the Great Recession.

Too often FSBO's and their hiring managers are

improperly focused on discrete skill sets versus trying to look across the gambit of how do you staff for a business model, delivery environment and industry foreign collaboration that is built on speed—rather than dogma. Taking cues from industry trends about the very consumer they are trying to serve and the individuals who have the necessary foundations to contribute, the landscape shifted why FSBO leaders believed the traditional boom-bust cycle would repeat itself—it won't.

A Deeper Dive: Yellow Lights

There are dozens more of these tactical and strategic uncertainties especially when we review the educational debts being carried by U.S. households. In a recent Bloomberg article ("Student debt in America has hit a record high"), student debt levels are now over $1.3 trillion, a doubling since 2009. Whereas total household debt has fallen by 1%—mortgage debt down by 8%, credit card down by 10% and home equity down 33%—student debt along with auto (up 42%) have increased with each now over $1 trillion USD.

The New York Federal Reserve, whose research underpinned the statistics for Bloomberg, also noted that one-quarter of student loans are 90 days past due, the government guarantees 90% of the student loan market, and next to credit cards, student loan repayments are often the least important for borrowers to repay. And when we factor in that corporate learning or training results at 20% to 30% retention, as compared to employees who self-learn (with retention over 70%), the traditional strategies and methods of onboarding personnel (who, what, where, and how) and expanding the skills of those already employed, begin to look suspect.

A Brain Drain—By Avoidance

In reviewing the last several years and forecasting the decades ahead, there are probably 50 different parameters that may need consideration mapped against a management practice—iteratively linked triple constraints. We have many trade-offs, and when profiled, FSBO's can start grouping items together.

FSBO's traditionally communicate and aid their new employees, but typically not overtly encouraging them to leverage independent skills and capabilities—they educate their people on conformance and lines of control. Within FSBO's everybody's got to conform the same way, often rationalized as a method to ensure regulatory compliance, and because FSBO's are big and geographically spread out. The implications of this rationale are when you offer diverse products and services to a heterogenous and geographically dispersed environment, to reduce risks and variations. FSBO's need to have people that are thinking the same, acting the same, and working towards the same processes—or potential face heavy investment losses and brand repercussions.

And while FSBO's have arrived at their BAU's to maximize profits while controlling risks, potential and current employees don't necessarily want to work what they perceive as self-serving organizations. They think of innovation. They grew up very differently, and so, if you have those types of skills, and you're in the top or upper half echelon in terms of skill sets and capabilities, that's not necessarily where they want to be. This mismatch of ideals versus cultures is going to be a huge challenge until FSBO's begin to realize that they need to align their BAU with the goals and priorities of the people they're trying to hire. These are the things that range from social impacts, to how do you provide empathy to the customer, to how do you deal with knowledge versus the intellect versus stability, and just a whole host of parameters which are fluid.

Therefore, FSBO's have seen such a drain, and why, in many cases, their staffing models have gone increasingly to outsourcing. Additionally, there have been rapid rises in flexible or dynamic collaboration models with FinTech firms who meet their customer needs and aid with market timings. For if you want the best talent, and you are an

FSBO who has negative market perceptions, you cannot simply hire potential candidates (or you will pay a high price for the person). Secondly, even if they could pay for it, those people don't necessarily want to work for FSBO's given the plethora of reservations stated in the previous chapters.

Moreover, these younger workers are taking things into account that traditional bankers typically have not, as conditions for employment—like the environment. For those within the financial industries they believe, "We're bankers, what does it matter to the deal or transaction?" Well it does matter to these younger demographics, and as it is about collaboration, impact to society and adding value. FSBO employment prospects view their contributions as more than a transaction and much more than a deal.

As we said, achieving alignment and an ability to create a flow of quality personnel starts with the business model. What are you really trying to do? Who are you trying to serve? If you have a business model that promotes a culture, which drives their salespeople to create fake accounts complete with fraudulent metrics and analytics, then what type of staffing and skills are required? If you're a potential employee who was being recruited to an institution, who only after regulators intervened to stop this organization's cultural practice (and this happened in 2016), why would you "waste" your talent on a "damaged" institution? We're not sure why you'd want to go into that environment if you have skill sets in demand.

If you (as a skilled worker) want to do something innovative, you wanted to do something that was game-changing or socially responsible, why would you join a brand name FSBO who has committed fraud and bring actions against their own customers. For the employee prospect, there thinking starts to form around, "Wow, do I really think these people are truthful? They may say one thing, but they're likely going to do something totally different, once I'm in the door."

So, it goes back to the culture shaped and managed by the very business model that many FSBO's proudly point to as what differentiates them within their markets. It has efficacy to me, you and anyone seeking a career—not just part of a transitory program of work. Experientially, it has always mattered to us as workers and those striving for career-long learning.

Any of us could always walk into a financial meeting, to a board meeting, we could give a presentation, and people would say, "Let's see where this goes. Perhaps you're right." However, when the meetings were over, repeated actions did not align with the words. You perform this monkey dance enough times, you create a self-fulfilling prophesy if you will, and people begin to say, "Is that really what I want to do?" And unless you're one of the Big-5 FSBO's that controls nearly 50% of the consumer markets, have the breadth and demand for solutions, how are you going to attract new talent to do continue to build the enterprise?

We are confident of the statistics showing that over one-third of the population in the U.S. hope that Amazon or Google (and other tech brands) are going to get into the banking business. If they enter these regulated industries (either as partners or with whitelabeled offerings), they're going to be innovative because they have the capability, they have the technology foundation to staff, they have the customer experience, but they don't have the business industry (hence, they may buy a beachhead via an M&A).

For large and complex global innovators, this represents a trade-off. Will they seek experts to build their own offerings using their delivery platforms in retail or information management to staff content experts? Are they going to seek out the best process experts? Are they looking for industry data experts? Time will tell, but if Amazon's purchase of Whole Foods is a framework for disintermediating commoditized industries, then it will be leveraged by others applying similar principles to disparate industry models[lxxxvi]--and the skill sets that follow their success.

FSBO's historically adhere to a cafeteria plan mindset (if we offer enough within a financial product suite we will get and retain customers). Oftentimes FSBO's don't question what they believe is an axiom—"We always did this from a business to consumer model." By not adapting to consumer and market tectonic shifts across a financially interconnected world, a peer-to-peer world, FSBO's cannot adequately achieve the synergies they need with a void of professional talent and commitment. That is something very different which is not only threatening for FSBO's, but they don't quite know what to do with it. It stems from their model not reflecting any reward compensation for participating in a world where their control is secondary to the offerings demanded by the end consumer—and hence they are experiencing a brain drain by avoidance.

A Deeper Filter: Traditional to New Realities

It is against this backdrop that FSBO's are seeking qualified borrowers—ones who are exponentially tech savvy and mobile delivery driven. Moreover, within FSBO operations, they are rapidly moving from a build to a consume model thereby significantly altering the staffing and organizational configuration necessary for these aged business models to succeed.

To shed additional focus of the complexities involved within these permanent shifts, the following diagram illustrates the transference from the old to the new, and the limitations and filters wedged in between their realization.

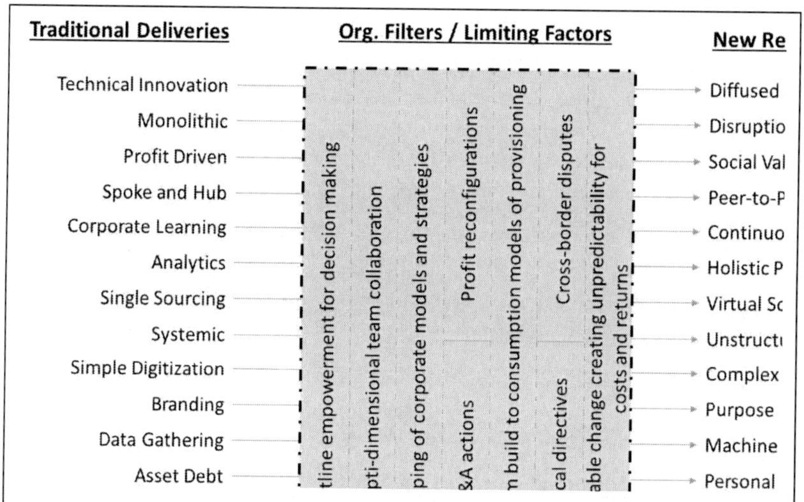

As can be deduced from this diagram, the path to a future model involves a journey that will unlikely be linear or a singular budget line-item. As FSBO's seek new staff, the recruitment methods and model's necessary to transform the enterprise will need substantial rework. Hiring managers familiar with seeking discrete skills and profiles will grasp for recruits and workers whose skills and aspirations reside outside the norm. Additionally, candidates don't view banks or lenders as partners who aid society let alone institutions who foster innovation.

Preparing for a New Reality

So, if we had to give the first steps for change and improvement for a FSBO, what are the items you'd start on right then? What should you do to ensure sustainable transformation across divisions to guarantee your FSBO becomes the place highly sought after, highly skilled personnel want to join, and collaborate on innovative and meaningful financial solutions?

Every transformation begins with an objective assessment of what capabilities are not only present within the FSBO and its partners, but also what is missing or potentially a critical need. For example, "If this is our current state, if this is where we're at today, this is where we need to be, and these are the types of skills and capabilities, which we have to have. What's the gap and how can it be bridged?" That's the traditional model that's been out there. We superficially know how to apply it to projects and their larger cousin's called programs (i.e., a collection of projects). Moreover, it becomes a remedy solving exercise to say that if these are our problems, then what are the solutions that we need to do? And then we can figure out can we do it, should we do it, can it even be done profitably?

That's the first step to determine what that gap is comprised of and how much of a leap will be required to move from today's reality towards tomorrow's promise. Additionally, this analysis provides a set of opportunities and options that, if you don't have them (i.e., the gap demands), then you're going to basically chase the Holy Grail resulting in lost time, ruined careers, and failed market launches.

Frequently, FSBO leadership condones only a cursory attempt at creating a baseline due to competitive factors and a belief that if they can do it, so can we. Yet, your FSBO doesn't really have a good idea of what personnel can be successful at with new releases and initiatives. Do existing personnel have the foundational literacy and skills to compete? How do they want to be trained, which is a whole different discussion that needs to be aligned with the cultural internalized beliefs? With this roadmap sketched out, your FSBO can start conducting learning and development programs. Finally, you start to say, "Okay, what skill sets are we really missing?" What skills sets do we need to hire or internalize? What do we do with joint ventures or with outsourcing?

So, when it comes to training, a lot of the diecast models are evolving or being discarded. We are in highly optimistic with this e-learning such as Udemy and Skillshare. We have even taken a MIT class through Coursera and found it not only improved our skills, but also meet the way we wanted to learn, while fitting our busy schedules. FSBO's traditionally, have not embraced these asynchronous approaches, and yet we're seeing more online training being offered by trade groups such as the American Bankers Association (ABA), the national Mortgage Bankers Association (MBA), and ALTA (American Land Title Association). Moreover, if we examine progressive colleges around the country from community colleges to four-year schools, the trends and models are clear and very advanced.

A Deeper Reality: Objectively Recognize the Limitations

The most striking limiting factors that must be first confronted when approaching the realities above is the existing organizational culture and structure (still operating under a command and control approach). These two core items, some might argue they are competencies in the largest institutions, create an identity for people within the enterprise, and over the decades has served the organizational profitability and brand identity well. However, this culture and structure are the starting points for sustainable change to occur and are often the greatest barriers for recruiting and retaining talent.

The U.S. Bureau of Labor Statistics recently published that Millennials (those under 38) stayed an average of 1.6 years in their job. As the new decade dawns, this age demographic will represent approximately 45% of the total workforce under 65, and with unemployment likely to remain between 4.5% and 6%, those seeking jobs will gravitate towards (and stay at) firms that deliver more than a traditional career path—more than the "banker's' way".

Including a pent-up demand for housing (due to loan availability for first time borrowers and 30 years low for starter homes), and with this same demographic being the least employment mobile in generations, FSBO's face substantial headwinds when it comes to securing personnel, contractors or even tightly coupled outsourcing arrangements (in the face of looming H1-B visa restrictions).

Traditional Learning Principles are Outdated

First and foremost, using our framework provided above is, "What do we have today?" The implication, as part of the principle of creating a baseline, requires an analysis for anything capable of reuse (e.g., skills, infrastructure, capabilities, centers of excellence, et al), given the business model. If we're going to train personnel and contractors, we better have a good idea what we need to augment, and what the process will be. This leads to the second realization, "Okay, that's what we have. How do people learn now days? What's the best approach to not only provide skills, but increase retention?"

For every functional category or bridge needed, we need to ask, "What do we need to do and how do we need to learn?" This simplistic query has been underpinned by studies from the large business leadership firms that confirmed most existing employees and contractors, a range of 60% to 75% when averaged out, say that really want to learn something new and will take risks. The other percentages fell within, “Hey, I got in here. I'm just going to ride this thing out."

However, in most cases, those people that really want to (re) learn and they want to invest in the FSBO—and they'll invest their own time, if they can do it the way they like to learn. Why? Because the interactive, video based or even hunt-and-peck learning strategies solo or within a small group mimic what many have become accustomed to doing socially. For FSBO leadership who have risen from time and grade to their current positions, this is a very atypical type of a non-classroom approach. We don't necessarily tape a lecture and then spew it out like a lot of these classic e-learning approaches saying, "This is what you got to do."

There is something to be said about peer-to-peer groups, anywhere between two to six or two to ten collaborative learning “pods”. From experiential feedback we have discovered, there is a positive result from the combination of everybody asking questions, sharing their experiences. Individuals, supported by experts in their field, leverage experiences and strengths that can be far more impactful than the less than 20% retention frequently achieved by traditional deliveries and interactions. It becomes a collaborative initiative and that also, then, establishes the baseline of the people that must work together. Who's got the skill sets? What do we really need to do? Can they get along and become better together rather than some just individual contributors?

If we're a group leader and we're embracing, "Let's go back to the days of high-performance teams. I have to get a tiger team ready assembled to roll out X, Y or Z to the market." Well, we're going to need all sorts of methods and techniques to get them to learn because with current FSBO operational requirements, we probably need to conduct this learning beyond the eight to five time clock. Why? Banks are built on day-to-day types of environments—FSBO's live and die by meeting or exceeding day-to-day. These people are fragmented by day-to-day activities, and in fact, their reward structure is directly tied to its achievement. They frequently don't have time to learn and they're not really focused on it anyway because they're worried about, "What's back at my desk?"

Therefore, you need to offer them an atypical type of learning process and that will involve some of the e-learning, that will involve the small team building and offer collaboration assistance to share experiences and build confidence. Then you begin to also open up your social media channels where people can dynamically collaborate online, offline, together to solve issues, to learn, and to get that experience and skills that they need. It's a different process, and one where those getting training and additional skills are not, "... simply checking-the-boxes."

A Deeper Trade-off: Achieving a Necessary Alignment

Ask yourself, how many enterprises seek top talent only to have them leave within the first year? What is done to ensure that employee / contractor onboarding meets the expectations for both parties? Why are enterprises using for leading-edge talent a model that is rooted in lifetime employment (i.e., climbing the ladder)? Career FSBO HR professionals can add a great many more challenges, but the bullseye is squarely focused on the organization and elaborate set of triple constraints that define and govern the enterprise.

A visual representation of the tradeoffs and controlling factors in action are illustrated in the following diagram.

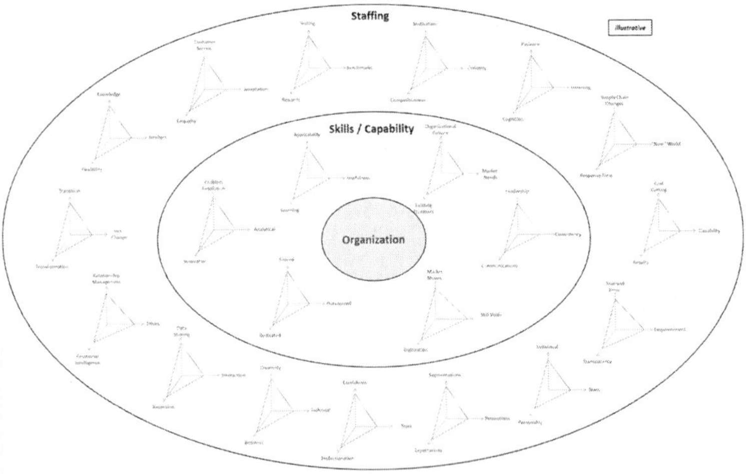

Also, demonstrated in the above diagram is that as FSBO's undergo permanent cost transformations brought forth by operating, technology and delivery changes, the weakest link across all these morphing segments resides with capable personnel necessary to enact and sustain change.

For most managers, their inability to grasp the multifaceted restraints brought forth under their conservative and regulated cultures is that with the movement to complex digitization and comprehensive changes in work spurred by omni-modal consumer delivery models, their staffing desires are not aligned with hiring realities.

Changing the "Block and Tackling"

Remember as FSBO's, we have due to extensive regulation created a conformance culture, becoming obsessed in checking-off-boxes (remember your last FDIC audit?). After we go to training most line managers will respond in some or all of these themes, "Well, did they check it off? Did they achieve this skill based on a standard test? Did they complete that chapter?" However, that is not necessarily the best way to learn in an operational type of environment. If you're trying to re-educate people to behave the same way and understand the culture they are functioning within, that's one principled purpose. If you're trying to get them to deliver an innovative product, a solution set in a dynamically and vastly changing environment, which is global, that's a whole different process.

This is where the application of e-learning, that's where your teams come into service even if they are virtual teams that coexist globally—they don't have to be in the same proximity just linked by a common purpose. Additionally, today we have the video capabilities (much of them increasingly interactive and as part of virtual reality learning) that we didn't have five or six years ago. This includes public domain options such as YouTube, internal multimedia, expert websites, as well as countless "how to" portals.

Most FSBO training operations (and the structure and goals of how they are rewarded) either internally or externally, are not set up for new era innovations and changing employee behaviors. They're set up for a couple of these things. Whatever the organizational title within the enterprise responsible for learning and integration within or across divisional boundaries, they must bring a plethora of varied methods together to address what they need and that goes back to the skills, and the gaps and the bridges. "What do we need to do and when do we need to make this happen?" Because if we have six weeks, classroom learning is never going to work unless all the individuals assigned have nothing else to do—and we know that goes for an e-classroom or a physical classroom.

Yes, it's just seems like these learning and training challenges are going to be more intense as banks roll out new products to meet highly specialized customer segments down to demographics that cover just a few hundred financial consumers. In total, it starts to sound a bit overwhelming—and it does.

Let's go back to what we said when we started this. What type of business and delivery model are we applying to the staffing, the training, the operations and organization? Are we're going into delivering (discrete) or transforming (broad) the FSBO, and communicating to our institution that we must own or control historically the solution? Are we going in saying, "We control the outsourcing arrangement and they work for us"? Or do we state that, “Only apps and code developed in-house will be accepted into production”—that is not a collaborative environment. We don’t care if it is labeled something else, it cannot be economically and brand risk acceptable where a FSBO is the master of all IT and process aspects moving forward. FSBO’s have permanently lost their grip and as such, they need to redesign their training methods and techniques to align with the new market realities.

What we're talking about now is we are going to build dynamic relationships and training modules to create a capability to compartmentalize, and assemble “stacks” of training that meets organizational needs. This leads into the next section on "How do we make this happen?" Banks have traditionally been focused on “hire locally” and they can create a matrix of skills to showcase depth—but not necessarily an ability to incorporates them into a program of work. Why? Because it never really has worked due to individual fiefdoms—like various government programs. Managers and program leaders are restrained by political silos who take exception to change and whose reward structures mirror the antiquated FSBO.

A Deeper Plunge: FSBO's are Not Your Father's

And finally, it is worth noting, that with the continual decrease in traditional banks (approximately 6,000 per the FDIC) and a reduction in physical branches projected to top 15%-25% in the next decade, the type of employee available is materially impacted by how as consumers they are using FBO's.

Factor in that corporations are estimated to be the largest holders of housing by 2025, and we can ascertain that for all the screening, testing and mentoring developed since 2008 is not going to serve a fundamental shift of delivery and strategy, which is taking place across FSBO's, and in particular, housing.

Therefore, as FSBO's assess their place within an evolving domestic and global financial supply chain, there will be permanent shifts in their business models. If those models change, so must the strategy not just for delivery, but also for bolstering up the weakest links of the supply chain—people, skills and an organization designed for a time gone-by.

Some will believe that AI or augmented reality implemented via a blackbox or open standard will provide technology that removes the reality of creating a highly capable and efficient organization. Perhaps, but like all fundamental transformations, they are not apps. They will require a substantial investment and strategy adjustment.
As FSBO's we've invested billions in blockchain, outsourcing, regulatory compliance, mobile solutions, marketing and lobbying. But as the political focus concentrates squarely on the workforces, student debts and a vision for the future, we may find, if we examine it

with objectivity, that our approaches for training, staffing, education, hiring and retention will miss their marks.

Rewards Breed Cultural Dogma—And Future Changes

For those FSBO's steeped in tradition, that's how we have always staffed even though we have experienced vernacular adjustments and deployed different techniques—top down. That's how we've trained—classroom and check-the-box. Think about it. If you're in a training module for let's say a new computer platform (i.e., process, apps, data, technology), people undergoing the information sharing and day-to-day adjustments don't usually take training to learn about the impact on the business segment or profitability.

For example, they're just told to deal with the transaction based systems, crosslinked data isolation modules, transaction routing and sales dates, and cybersecurity. That's what they're training. They're not necessarily training to understand how "this stuff" works or why it should be important to them. It's hard to internalize information when it has no context and cannot be openly discussed between participants.

However, it's the organization that is imposing a limit on what the training is and what the staffing models are. Therefore, we approached this chapter utilizing a different mindset. We approached it from the bottom, finally getting to the point of saying, "We can train in all sorts of these different ways from mobile phones to encapsulated apps, to taking tests online, to collaborating with other individuals around the world, and getting that information.

The correlation proof is out there through academics, through your own internal departments, through a third-party (e.g. outsourcers), through even competitive information, and through partner's disclosures, joint ventures and even failed launches. They all provide you that training compositions and results, which can be used as starting points for bridging gaps or making a conscience decision to abandon a market. Yet FSBO leadership often, with great bravado, says “We’re going to make this happen!” That's where the FSBO foundations begin to crumble, and yet another real reason why people are not eager to join financial services firms in general.

Yes, there are some big pay-offs if you become partner or very senior executive in a FSBO where you are paid vast sums in a base and achieve multimillion dollar performance bonuses. However, that's not the norm and potential employees and partners are beginning to say, "I'll be trapped in a box if I join this FSBO. I can't necessarily run up the line and figure out who has the final decision-making authority. I can see obviously that the enterprise is going to be profitable, but if join this operation my ability to contribute will be throttled by organizational design, regulatory guidance and a culture not conducive to being changed. I have a business view, I have this IT skill set, I am a good team player, but no matter what I have, it becomes ‘like fighting city hall’”. It becomes, an overwhelming process by which, “I can't make any progress, so why bother?”

If the organization is not aligned to be dynamically driven and aligned, to be results driven, then the FSBO is indirectly saying to the markets and prospects, “we hear you, but we change very slowly regardless of what the markets do.” (Ask yourself, how many times have you heard the phrase, “We can’t turn on a dime!”?) For someone considering the FSBO given the existing animosity and commoditization of its offerings, we're not doing the right things for the customer, we're not providing service, and we sure can’t delivery profitability. It becomes analogous to the firm above who committed fraud against its own customers.

FSBO leadership says, “This (fill in the blank) is the golden ring, this is what you need to do”, and everybody went to that because the organization rewarded those who performed accordingly (e.g., conformance). The reward structure, the analytics, everything was wrapped around how that business unit and sub-segment performed set by objective measures often conflicting with what was needed as part of the rollout. If the FSBO is still functioning within that mentality, it will be nearly impossible to move the business model to align it with education, staffing, and training as the organizational dynamics run counter to unrecognized success criteria.

Within FSBO circles, we talk about branches and how's what purpose and demographics they will serve in the future from how customers arrive, to what they use to self-serve kiosks and smartphone integrations (think Home Depot giving you aisle and bin locations for what you wish to purchase on your smartphone app). However, all this dialogue and frequent “window dressing” may not achieve the results spearheaded by the performance metrics and (hopefully there are) baselines.

Moreover, across FSBO’s leadership ranks, we drive products and services by organizational unit, and that is the wrong thing to do for the right reasons. It's a false structure by which you try to get results. We've seen that it doesn't work and it doesn't drive innovation, and it doesn't provide the skill sets you need, and it doesn't provide the right set of staff mix you need to make that happen. It’s an artificial construct hemmed by politics and fiefdoms.

A dynamically driven institution, which is what we're talking about, can be seen functionally in the previous callout box within the diagram. It is a balance between and tradeoffs among a host of variables continually being readjusted and reassembled to meet the demands of varied departments and leadership. Every challenge that we talk about, every solution you're going to bring up, is all talking about fundamentally restructuring the institution for a new marketplace.

Think of it as revealing the mysterious individual behind the magical curtain in the Wizard of Oz or Harry Potter with the cloak of invisibility. Most personnel within a FSBO believes their leaders are special, magical, and born to govern (sometimes with an iron fist). The reality for the new FSBO, which is being attacked on all fronts, is that these mystical, magical beings are fantasy. Most of these individuals have limited capabilities regarding decisions of how to truly innovate, and find the dynamically shifting business model that will lead to profitability and (organic and acquired) growth. Our "reverence" and fear of how these organizations operate is precisely why 80% to 95% of all budgets are being allocated to BAU.

It is this reality, which surfaced after years of work within FSBO's, aligning all the pieces, parts of solutions we assembled over the years, to arrive at these directions as a number one priority before embarking on chasing competitive pressures. We believe that one of the fundamental reasons why these deficiencies or voids have happened can be found in chapter six. Digitization is what drove efficiency in FSBO's for the last decade. This is rudimentary digitization coupled with the idea of big data, has brought us to a precipitous of future upheaval among many who dismissed these changes as ever being "that important".

For FSBO's their world has gone from simple models to complex, to multi-varied analysis, and multiple dimensions—that is where we are today. FSBO's are just not ready for this dimension as their control has not been challenged to the extensive range currently on-going and being projected. Moreover, it's because the organizations really don't know how to make this leap, so many ignore its manifestations and practice incremental change which guided them previously.

A Deeper Reality: The Next Wave of Outsourcing

FSBO's must become project managers or conductors comprehensively integrating all aspects of multiple outsourced arrangements. A familiar analogy might be a program office operated utilizing defined architectural filters deciding how and what gets implemented, but using external operations to carry out the goals, objectives and implementations. For FSBO's they retain the most critical components of IT and BP delivery in architecture, integration, and data, while releasing the burden of development and daily-operations (which can be shared if desired).

Moreover, using emerging outsourcing models, FSBO's can satisfy consumer, regulatory and business requirements faster and with a reduced spend. How? By accepting that core competencies have shifted. By recognizing that an institution can no longer create all its point-of-entry channels. By recognizing that FinTech is a by-product of regulatory environments, and a massive, emerging shift of consumer behaviors is on-going. By allowing partners to play a leadership role in delivery. By admitting that the FSBO industry models are fragmenting providing a unique opportunity to change the value equation of how we spend our budgets, and for what risk-reward returns.

The deployment of layered outsourcing models will be for FSBO's about being persistent and adaptive. Some, like the researcher seemingly producing materials to continue their business model, will say FSBO's don't have the expertise or track record to take on these unfamiliar operational roles. Yet, saying it doesn't make it so. FSBO's are operating systems, delivery channels, compliance requirements, legacy systems and large voids in skills already. The sophistication to manage effectively and efficiently layered outsourcing is no more than trying to balance IT budgets that are breaking all in an effort to stay operational.

What is the Recommendation? FSBO's cannot continue to advance with a very significant and draining BAU legacy (i.e., doing things like what they have done for years). FSBO's need to adopt the new layered approaches to outsourcing (e.g., orchestrated, networked), and take a centrist view of architecture, while partnering with many likely outsource providers both domestically and internationally.

Additionally, find partners (i.e., traditional outsourcers, vendors, consultants) who can provide not only a sounding board for your directions, but who are willing to provide unbiased, external "...did you think of ..."—even if it is not what they offer. Finally, accept that market movements and consumer trends will not be measured in years moving forward—but in months. The markets and consumers will continue to be fluid and likely accelerate delivery demands, while decreasing solution longevity.

Those of us who started the ITO and BPO efforts over two decades ago are shaking our heads in wonder at those who vehemently describe the state-of-the-industry with models that were already antiquated nearly 10 years past. For those FSBO's leaders who are seeking a profitable way forward, the idea of layered outsourcing should hold immense interest and potential—as it provides means to address key regulatory areas of protection, transparency, security and adaption (e.g., stress).

Start at the Organization and Work Backwards

The contents of this chapter may be distasteful or even heretical for people who have spent their entire careers within FSBO's. But when you peel back the layers of "onion" created by Davos, by executive benefit packages, by government bailouts and buybacks, by fake accounts, by outright fraud, and by less-than humble FSBO personalities, we now begin to comprehend why individual consumers are so thrilled with idea of Amazon and Google being part of the FSBO world. For consumers, they understand how well that information fits, promote strong loyalty across their offerings and invest heavily in their personnel making these "new bellwethers" consistently top in places to work.

It has been the breadth of digitization which has created a void for most FSBO's when it comes to clearly understand their customers and securing the personnel or skill sets necessary to compete and gain the attention of those digitally savvy, and highly profitable demographics.

Starting from today, what are the drivers that are causing our future FSBO models to succeed or fail? What do we even need to alter or extend our positive results? This is where gaps and bridges come into focus. These disciplines and methods of change have been around for a lot of years and will be around for many years to come as they adopt to FSBO environments. However, if we only look at it from an organizational perspective, then oftentimes that's why we get so segmented within FSBO's. We have all these layers and silos because that's all we ever look at. We often only look at our own little wheelhouse and we repurpose and project from what we are comfortable.

And that's the problem. That's certainly what we do. We turn the nut. However, there's a whole FSBO being transformed and built. We're turning one nut.

And the changes and demands in staffing are only going to move faster with the rapid advances in VR, complex data integration, machine learning, algorithmic matching of customers to capabilities. Factor in peer-to-peer processing, settlements, how do we do loans, and this has completely changed the mix and the people, training, and organization needed to leverage these advancements. It truly comes down to the dynamics of how we interact. How we lend, how we take things back, how we socially interact, how do we even track information, and how do we determine creditworthiness. That's all changed and is still evolving. But our actions and future results starts at the organization and moves backwards.

Illustrative Challenges / Opportunities

Category	Illustrative Challenges / Opportunities	Controls and Implications (The 4-V's) Volume/ Rate	Variety /Adapt	Velocity/ Freq.	Veracity/ Accuracy
1	Recognize and implement staff and training programs that put more control into the hands of employees and partners linked to performance results.	↑	↑	↑	↑
1	Redesign and iterate organizational structures to accommodate change, while eliminating the largest obstacles to collaboration and profits.	↑	↑	↑	↑
1	Design and implement organizational filters that positively contribute to "new reality" achievement.	↑	↑	↑	↑
1	Creating retention programs to incentivize employees across a critical first year of onboarding.	o	↑	↑	↑

Cate-gory	Illustrative Challenges / Opportunities	Controls and Implications (The 4-V's) Volume/ Rate	Variety /Adapt	Velocity/ Freq.	Veracity/ Accuracy
2	Deploy and measure trends and results associated with continuous transformation from current to future states with transitory skill sets mapped out.	O	↑	↑	↑
2	Create and capture data components to accurately trend and project staffing, organizational design, and training across states of transformation—not after the fact.	O	↑	↑	↑
3	Design learning programs for multiple models of retention as well as critical mastery of necessary capabilities.	O	↑	↑	O
3	Reduction of check-the-box screening to meet an increasingly subjective hiring model that moves beyond simplex profiles.	O	↑	O	↑

↑ – Increasing non-linearly, unpredictable behaviors
O– Increasing but predictable

NOTES

"My question, Moses: What is the threat of new entrants into this market?"

Image ©Shutterstock, Inc.

8 Marketing

"If you create a purpose, make sure you deliver on that purpose. It is the filter through which you make every decision." – **Anne Finucane**, Global Chief Strategy and Marketing Officer, Bank of America.

FSBO marketing executives are currently dealing with a problem that marketers in other industries rarely face. Under normal conditions, the marketing department seeks out likely prospects for the sales team, often creating buyer personas to represent consumers most likely to require the company's products or services. To attract prospects, the marketing team proposes adjustments to the pricing, the promotion, the placement, and occasionally even the product itself until a suitably large universe of prospects can be delivered to the sales team.

But what happens when the reason likely prospects shun the company's offering has nothing to do with the product, pricing, promotion or placement? What happens when the reason it's difficult to attract prospects is that consumers have lost respect for your industry?

Modern marketers have added a fifth P to the traditional marketing mix: people. It's the relationships between the company's people and its customers, whether real or, as in the case of a celebrity spokesperson, imagined that lead to sales. Those relationships are the result of many conversations, often about the other four P's of marketing. The challenge for FSBO's is that they do not have the people who know how to create meaningful relationships with the new breed of banking customer.

Worse than this, the traditional FSBO model has placed the key banking relationships between the FSBO personnel and a group of FSBO non-customers made up of liquidity sources and regulators. The actual FSBO customer has traditionally been treated as a work input, providing deposits[lxxxvii] or promissory notes that can be leveraged for FSBO profits.

Part of the reason FSBO's have been distracted from this problem is that technology has been changing so quickly that most have assumed that FSBO's have a technology problem instead of a customer relationships problem.

In truth, part of the reason that consumers began to take their business to FinTech firms was that FSBO's had refused to make the digital transformations[lxxxviii] required to deliver their offerings online. That has largely changed now and most FSBO boards' understand the importance of making the digital transformation. However, that will not be enough to attract consumers in the future. FSBO's have lost the first mover advantage for online consumer financial management, and are left with a dwindling client base of consumers who are wondering why they really need a traditional FSBO anyway.

To be effective, FSBO marketers will have to rebuild trust on the part of the consumer, draw clear connections between FSBO products and consumer needs, and then become expert at starting and maintaining substantive conversations with FSBO customers, which will build the relationships that lead to sales. These are exactly the tactics that FinTech firms are using today to steal FSBO customers away and achieve outstanding market valuations.

FSBO marketing done well will respond effectively to three of the risks we have identified in this book. It will serve as a counter to market animosity while rebuilding trust. It will slow disintermediation by revealing the FSBO as an option equal to or better than their FinTech competitors. It will beat back commoditization by creating personalized connections between the FSBO and its customers.

To accomplish all of this, the FSBO must redefine its relationship with its customer base. That starts with ensuring that the institution is marketing to the right consumers in the first place.

What Comes First, the Product or the Customer?

In the beginning, FSBO's took investor money and loaned it out with interest. From the beginning, the liquidity source was far more important than the customer to whom the FSBO would loan that money. Depository institutions made the FSBO customer a source of liquidity, but even then, the customer base was segmented by the size of their deposit accounts, with the largest FSBO customers receiving preferential treatment and the rest feeding at the same trough of FSBO products and services.

It didn't make as much sense to segment the customer base by the size of the customer liability, as the higher the loan balance the higher the risk. Better to serve more customers of moderate to low risk with similar, easy to originate FSBO products, and so FSBO's began walking the path to commoditization by their marketing efforts early on.

That's not the way FinTech firms entered the market. On the surface, it appears they simply built a financial services app and pushed it out through the various app marketplaces, taking every customer at one of finite number of price points (usually 3 to 5 tiers, based on anticipated usage or included functionality). That is what you'll see if you visit one of their landing pages, but that's just one part of their marketing system.

FinTech firms grew out of the mobile app boom that itself was a result of the trend of consumers spending more time online. Consumers didn't go online initially to buy products or stream movies or apply for a FSBO product.

They went there to do research on products and services and to be social. It was social media that pulled Americans online, in the form of millions of AOL CD-ROMs mailed to them relentlessly. From the very beginning, the audience of consumers was the most important part of the process, in stark contrast to traditional financial services.

To make matters exponentially worse for FSBO's, the home mortgage foreclosure crisis convinced millions of American homeowners—and their growing children—that it was the note that mattered to the FSBO, not the borrower. It doesn't help to point out that FSBO's were driven to foreclose by investors and that failing to do so efficiently was the most destructive force American cities had faced since the Great Depression. The decimation of cities like Chicago, Detroit and Atlanta (and many more) and the zombie foreclosure crisis were two direct results. It wasn't that foreclose was an unnecessary evil. It was necessary. What wasn't necessary was the way consumers were treated during the crisis, while their children watched helplessly.

If FSBO's hope to protect their traditional businesses from the new wave of competitors, they must change the way they think about and treat the banking customer.

A Deeper Dive: FSBO's and Social Media

Back in 2010, one of us wrote a column in HousingWire that suggested that "the mortgage banking industry is finally moving toward the social media tools that other industries have been finding so useful over the past couple of years. I know this because I can hear them kicking and screaming."

We were a few years too early on that call. Mortgage lenders and bankers in general soon found that compliance concerns would force them to put excessive controls on the social media activity of their sales people and loan officers. This would put them at a disadvantage that would not become clear until the Fintech firms began to easily chip away at the bank's market share.

Eventually, institutions found that the right social media policy could empower their teams to interact with prospective borrowers in a compliant manner. The next hurdle was overcoming the one-way communication model that has dominated the industry since the beginning.

Social media is a two-way communication medium. The old methods of telling consumers what they should do and why have given way to the new method of showing consumers something they can't live without and then making it easy for them to get it. This is what Fintech firms do, while banks are still telling consumers what they must do to ensure bank compliance. That's a failing strategy.

If two-way communication isn't something the bank is willing to engage in at present, and that's understandable given the bile they are likely to receive through any open channel, at least consider reversing the traditional direction of your one-way communication. Social media can be an incredible resource for an institution that is willing to listen to the voice of the customer.

Failing that, start following the bank's Fintech competitors. Watching how they use social media to engage with prospects and their user base will be educational.

Finding the Modern Financial Services Customer

It could be argued that the only reason a FSBO customer ever existed was because there was no other place the consumer could access bank products and services. That's not true today. While a banker may consider any consumer, who wanders into a branch a customer, it's highly unlikely that the consumer does. In fact, a study by

Goldman Sachs revealed that one third of millennials between the ages of 21 and 38 don't feel that having a relationship with a FSBO is something they will need in the future.

Even before the rise of FinTech firms, competition between FSBO's educated consumers that FSBO products between institutions were largely similar. They chose checking accounts based on fees and mortgages on rates. FSBO's, for their part, advertised those consumer decision criteria relentlessly, speeding their progress down the road to commoditization.

Reconnecting the traditional FSBO's products and services to the perceived needs of the consumer is an important step in the evolution of FSBO marketing. The need to store, borrow, manage, and profit from capital remains a core need of the FSBO customer. Unfortunately, there are now a great many competitors vying for this business, many of whom have first mover advantage into the digital space. While these services are the price of entry for the modern financial services institution, they do not constitute the battlefield[lxxxix] upon which the FSBO will win new clients.

The doorway into a relationship with the new FSBO customer will involve filling a customer need that goes beyond those filled by traditional banking products even when they involve those products. For instance, there are many new FinTech firms (and FSBO's partnering with such firms) that can offer an 18-year-old new bank customer who was born in America with a bank account. He punches a few buttons on his cell phone and he's set up. The same cannot be said of a foreign-born university professor in the country on a three-year visa and subject to all the rules of the Patriot Act.

The very wealthy is another niche the traditional FSBO can serve better if they choose to segment these customers by need instead of portfolio size. Their needs are different than other FSBO customers. Fortunately, this is an area in which many FSBO's have already been working to build relationships. For some, it's the only segment of their client base they serve with "relationship banking." That will change.

Business banking is poised to get bigger for FSBO's. This segment of customers has the potential to make up a significant part of the FSBO's customer base. And not just the big companies. There are currently over 28 million solopreneurs the IRS is tracking. These are self-employed individuals that have no employees but experience significant cash flow and require many banking services. Traditionally, small business lending has been picked up or dropped depending upon the FSBO's need for additional review. That won't work in the future.

Currently, there are many FinTech firm's intent on stealing away the traditional FSBO's small business customer base. Companies like PayPal and Square are filling the gaps banks have left and filling needs that FSBO's are ignoring. As those relationships develop, it will become increasingly difficult for FSBO's to win them back.

From emerging markets to industry verticals, there are many niches the traditional FSBO can exploit for customers, but only if they return to a more traditional approach to marketing and focus first on the needs of the banking customer.

Successful FSBO's will begin to segment their customer base by consumer need, not by bank product, and seek out those customers that cannot easily be serviced by the new, one-size-fits-all online offerings. In short, FSBO's must stop marketing commodity FSBO products and move to more traditional marketing, through which actual customer needs are discovered. By segmenting these customers by need and then wiring up existing or new bank products to those needs, FSBO's will be able to grow their customer base.

A Deeper Dive: Avoiding Marketing Metric Overload

The rise of social media and internet marketing has led to a new division in the bank's traditional marketing department. Digital marketing, a.k.a. Internet marketing or social media marketing, has quickly become as important to the FSBO as traditional advertising and media outreach. Because these marketing professionals work in the digital realm it is possible to more closely track their efforts and the corresponding results.

Because of all of this, tech firms have sprung up to provide data gathering and analytical tools that promise to give the professional marketers an overwhelming amount of information on every aspect of their programs. There are a number of risks inherent in investing in such tools.

First, there are only a few marketing metrics that matter to company management. These include marketing costs, marketing results (sales), reach and conversion rates. The new tools allow us to track a great many things that while possibly important for targeting marketing efforts, i.e., are we reaching the people we need to reach, don't generally tell us anything other than that the marketing department is very, very busy. Figuring out what data matters takes a lot of time and if you get it wrong you'll be watching the wrong door when the customer walks in.

This leads to the risk of management data overload. To justify budgets, marketing managers can dump buckets full of data on company management, little if any of it telling management what's being done right and what needs to change.

In our experience, the better designed the marketing campaign, the fewer metrics need to be tracked to determine success. The very best campaigns, in our minds, are those that drive prospects to a landing page, offer them a download and then wires to the company CRM so that sales know that the person placing an incoming downloaded the information. Three metrics.

Everything that must be done to drive prospects to the landing page should be tracked by the marketing department but is virtually useless to management, which should only be concerned with the costs and the results. If the expected lifetime value of the prospects coming into the pipeline for a new loan product or signing up for an account does not exceed the costs of the campaign, marketing is doing something wrong.

While tech firms would love FSBOs to spend a great deal of money on analytical tools, it's not that expensive to track the few key metrics that management needs to evaluable the marketing department's results. And spending less means spending less time sifting through piles of data in search of the few metrics that will tell whether the company's communications outreach is reaching customers or falling on deaf ears.

A New Approach to Creating FSBO Products

None of this is written to suggest that product development or a focus on good banking products will not still be vitally important in the future. The difference will be that instead of waiting for liquidity sources to offer FSBO's products to sell to their client bases, FSBO's must be using modern marketing methods to determine the actual needs of consumers, and then create products specifically to meet those needs.

The mortgage business offers a perfect example. Even though the home finance process is the most complex and least understood (by consumers) transaction most consumers will ever take part in, very little time or effort was taken on the part of mortgage lenders to make the process easier, clearer, or less stressful for borrowers. Lenders gave prospective borrowers information on interest rates and very occasionally closing costs. The rest of the process involved a flood of documents that consumers could read if they wished, but must sign without changes if they wanted the process to continue.

Many borrowers left the closing room feeling as if they had been taken somehow. After the foreclosure crisis, millions felt they had evidence to prove it.

Contrast this with the way most FinTech firms entered the banking space: first with a free and helpful app, backed by a flood of consumer-facing information about how to use the free tool to meet a real consumer need. Is it any wonder that consumers are ready to trust any firm that drops an app into Apple's online marketplace, but still believe that bankers are in the business of stealing their assets?

The first need of the consumer is information. Always. FSBO's should know this because for years—no, nearly two decades now—consumers have first gone online for information about their next mortgage loan. Instead of offering the information required to make consumers feel comfortable with the deal and understand the risks involved in the transaction, they perceived that the quickest path to close was to provide rate information, beat competitors with the lowest rate and the promise to "take care" of the customer and then put them through the same process they had been using for years (later speeded up by automation, but operating along the same process lines as they always had).

Today, we do see more FSBO's offering detailed information about the mortgage loan process, but lying as it does in the shadow of the foreclosure crisis[xc], it carries no credibility with consumers. They would surely believe information they read on a real estate site like Zillow.com than what is offered by a bank that hopes to originate their next mortgage.

The same analysis can be applied to any other product the FSBO might choose to offer. Today, cryptocurrencies[xci] are hot (see callout below). Will FSBO customers opt to do business in Bitcoin? Some of the largest FSBO's are betting they will and have already invested in blockchain technology to get a foothold in the market. Have they invested in developing the educational resources that will pull consumers to their offering, or are they following the traditional FSBO marketing script and expecting consumers to simply gravitate to them because they want the offering and the FSBO has spent millions on its brand? We promise you, consumers won't.

Consumers want information first and FSBO's need information from their consumers to find out what products to offer and how the consumer expects to receive them. Consumer information is a byproduct of conversations with the customer. As a happy coincidence, those conversations are also the key to relationship building, which the FSBO must engage in to succeed. This is the role of the new marketing department within the FSBO. Today and tomorrow, it's all about FSBO customer relationship marketing.

A Deeper Reality: Communicating Through Website Banners

Just when we thought banner advertising on websites was dead, retargeting allowed brand managers to follow customers as they traversed the web. Now, brand managers can put their banner in front of prospects on a wide variety of websites with a single ad buy. But getting them to click is still a challenge.

Here are some rules for turning website banners into a block of content the consumer doesn't even see into magnets for new business.

1. Never place an ad without baiting your hook. Advertising for "branding" is dead. Today, it's all about enticing your prospects back to your website or onto your mailing list. To do that, you need something to trade. Before you buy an ad, know what your prospects are going to be coming to get.

2. Build your ad campaign around your strongest value

proposition. Why are you a good value for your prospect's money? What can they get from you that they simply can't get anywhere else? What do you know that they don't? Think it through and use this as the core message for your advertisement.

3. Think about how your customer receives your value. Sometimes it's as easy as buying your product. Other times, your customers don't really get the full value until you add consulting, implementation, training and more. Make a list of all the things that go into making good on your value proposition.

4. Take one of the things on your list in No. 3 and give it away. Maybe its information (consulting), configuration tips or a list of existing interfaces (implementation) or a free training course or white paper, but get your prospects started on the journey to ownership by giving them a little bit of value up front.

5. Make them give you something for that value. Set up your website to get contact information from anyone who takes you up on your free offer. This is your next set of hot leads and it's the only thing that makes display advertising or e-mail marketing worth the money.

Building the Relationships that Lead to Sales

How will FSBO's build the kind of relationships that will repair their reputation and rebuilt client trust? By following the marketing path, laid down by Apple, Google, Amazon and the other leaders in the online world. Banking, like just about every other services consumers are buying today, must be digital first.

FSBO's will not have to start at the beginning. There

will be no burning and sending of CD-ROMs. But like the other online giants, FSBO's will need to find a way to become more tightly integrated with their target customers daily lives. In the process, they will be held to the same high standards that have been established by the online world's leading firms, even when they have nothing to do with financial services (though, invariably, they all do to some degree).

No longer will it be feasible for the FSBO to wait for a consumer to wander into a branch before beginning the work of winning share of wallet. The FSBO must begin to treat the consumer like a valued partner even before that consumer has spent a single dime with the FSBO. This is exactly what leaders in the online world have done in the past and what the new FinTech firms are doing today.

Cross-sell, upsell and share of wallet will still be important, but these will be tasks taken on only after the FSBO has a business relationship with the customer. They need to have a virtual relationship with the customer before that can happen. As we've already pointed out, building that relationship is a matter of engaging the customer in meaningful conversations built around information the customer perceives they need.

This should not be as difficult as FSBO's have made it in the past. It has been so difficult because it is very difficult to build a relationship with a party that is important to the transaction but not important to the institution. If one account holder is as good as another or one mortgage loan borrower is as good as another, FSBO's will struggle.

It's not that consumers want to be your only priority or your only customer. They just want to feel like they are. When a consumer opens their iTunes account or their Amazon Prime account of their Netflix account, this is how they feel. This is the new standard and FSBO marketing departments must be restructured to deliver it.

A Deeper Dive: The Competitive Advantage Community Banks can Defend

Fintech firms have outflanked traditional FSBOs by contacting consumers where they were spending time online and offering them apps that made financial services easier and more rewarding for them. There is no inherent barrier there that banks cannot easily overcome, which is what most of this book is about.

Meanwhile, banks, especially community banks, have a competitive advantage that Fintech firms will find very difficult to overcome if only the banks will capitalize on it. A Fintech firm based on another continent may be able to take customers from a bank by offering a financial tool that the bank cannot, but they will never be able to make an impact on the local community that a bank could if it chose to do so.

There are many ways for FSBOs to interact with their community that don't involve a business transaction. When handled correctly community involvement can lead to increased business. Here are some ideas we recommend for raising your company's profile in the community.

<u>VOLUNTEERISM</u>

Individuals: Whether Race for a Cure, Relay for Life or a Bowl-a-thon for Big Bros/Sis, there are plenty of places that someone in your company participates. This can simply be recognition of that time or a sanctioned event where the company supplies a team of workers at a Habitat for Humanity on a Saturday. This can be augmented by permission to dress down on Friday before the event, a Golf Shirt given to the team, or lunch being provided at the event.

Corporate: A corporate team can be created for a cause

because it is high profile in your community or because it means something to one of your departments. (Perhaps you have an employee that is battling MS or you lost an employee to ALS).

Participation in a Make a Difference Day: Have an organization team for a local or national Make a Difference Day. This may or may not take a day of work away depending whether it is scheduled on a weekday vs the weekend. It can be a press release from the company but could also be picked up by the local newspapers. This allows for good press to come out from a source other than from within your organization.

SPONSORSHIP

Music Festivals: These can be professional Music Festivals OR High School Competitions. The awards are often posted on the HS Websites and linked to various groups from the departments. Parents tend to look favorably toward the company that helps their children by sponsoring the awards or placing an ad in the program.

Program /Handouts: It is easy enough to create a nice business card sized ad as well as a ½ and ¼ page ad for things like yearbooks and programs. A photo of your award-winning building for a program of house tours that the Historical and Genealogy Groups are putting on gets your name in front of small banks and architects and realtors. HS Drama, HS Sports, and Dance Studios are great ways to get your name in front of parents. Of course, Community Theatre is more often patronized by those who have registered to vote and or have attended college. These run from 25 -500 dollars.

Community Festivals: These can be a great way for your name to be out there for several weeks at a time as they generally have advertisements running for longer periods of time (they are PSA's). Your name becomes linked with benevolence. This is a great way to get a company exec.'s face out there in a positive light.

GIVE AWARDS

Art Awards: Sponsorship of an art award or show can be as simple as being a photo contest or as complicated as a multi medium contest. You can have the schools display the entries and have the winner become a display in your offices. It can be youth based for different levels of school children or open to the community. If you have a place to display the art, you can have an artwork reception one evening to allow the community in. This allows you to educate the community as to what you do.

Poster Contests: Sponsorship of a poster contest can be aimed right at parents through their children. You could partner with a company with whom you are interested in working. It allows you to add an education component such as Lending and Fraud or whatever topic you are interested in focusing on.

HS or College Scholarships or Awards: Sponsor a $100 scholarship at several schools in your county or one school in your town or your local community college. This allows you to reward a student that has worked in a certain industry or has excelled in a school subject or has exemplary test scores. You can use this, for example, to call attention to the number one student in Business and Accounting or in Finance. You can also sponsor a student of the Month each month during the school year for the cost of a plaque and breakfast.

SEMINARS

Adults: Seminars about credit or loans can be given one or two days a year. This could include education through

Community Centers, Libraries, Senior Centers or Family First for adults that receive TANF (Temporary Assistance for Needy Families) through federal funding in a Family First or EARN program at your County Assistance Office.

Youth: Seminars can be given for Scouts for Financial badges or to Students. Host a CEO for a Day event for your local schools' Student Leaders.

FOLLOWUP

Whatever you choose to do, make sure you make the most of it with a well-planned communications effort that starts well before the event and keeps the experience in the minds of your prospects long after it's over.

Restructuring the Marketing Department

Chief Marketing Officers (CMO) in FSBO's must become more like their counterparts in other consumer tech industries and less like heads of production for various bank products. Instead of being tasked with ensuring the sale of more FSBO products, they must make their most important metric of success the FSBO customer relationship.

As they shift their focus, they will begin to build marketing departments that look much more like those at the leading online firms and less like what the traditional FSBO has used in the past. Every change will be made as a direct result of the new priorities of the FSBO marketing department, starting with the way the department is structured.

The newspaper advertising and newsletter marketing departments may remain, but will no longer be the primary tools the FSBO uses to attract new customers. Instead, of marketing impressions, the CMO[xcii] will go in search of online engagement. Using a range of marketing channels, FSBO marketers will hunt consumers online and snare them with valuable information, pulling them into conversations in the hope of building relationships that lead to sales.

FSBO's must learn to value consumer relationships beyond the value of the products the FSBO customer consumes. In truth, the value of any customer is far more than that customer's personal business with the FSBO, if the FSBO can leverage that customer into new customer relationships. In a fully connected online environment, each customer is the gateway to their own social network, and the FSBO's marketing department must be tasked with breaching that portal and winning that business.

Because of this new priority, the metrics used to track marketing department success must be changed from number of commodities sold to number and value of consumer relationships. This means that the marketing department of the future must have both a strong research department and access to powerful analytics. Instead of looking like an extension of the FSBO's communication or production department, the marketing department will begin to look more like an extension of the FSBO's network operations center.

The cramped, paper-filled office in the basement that previously held the small team of marketers who created newspaper advertisements and printed out FSBO newsletters will give way to sophisticated control rooms that track FSBO customer interactions in real time, foster two-way communication with those customers and entice them to share the FSBO's brand with their social networks.

At the same time, this high-tech marketing information center will fuel the FSBO's efforts to find, create or coopt online conversations with prospective customers. Once a part of those conversations, the relationship building process can begin.

Finally, instead of simply being a tool for marketing the FSBO's current menu of products or services, the marketing department of the future FSBO will promote the

FSBO's brand as a community partner. Just like the most successful nonprofit organizations, the CMO will guide the conversation away from the specifics of FSBO products that are too complex to describe in a soundbite and toward issues that impact the community and the FSBO customers living within it. In doing so, the FSBO's mission will be revealed as a mandate to build a better community instead of an operation that delivers a certain volume of financial instruments into the secondary market.

Millennials will embrace these FSBO's, fueling a resurgence in the local branch at the same time they engage online with an institution they believe is actively engaged in doing something they have internally denoted as "good."

Naturally, this approach will likely benefit the community FSBO or local credit union more than large national banking brands. As these institutions are more likely to make a strong case for social value alignment with local prospects. They can build a significant marketing advantage out of the concept of being the community bank that is in the business of building the community?

For the larger financial institution with more resources to spend on researching consumer needs and then providing valuable information in more places online, their advantage will be a direct result of the value they place on aligning their offerings with the consumer needs they discover in the marketplace.

In every case, the new FSBO marketing department will focus its efforts and gage its success by the number and depth of customer relationships it develops. The product marketing of the past that seemed so important to FSBO marketing executives will then largely take care of itself. Failure to re-prioritize FSBO marketing efforts will prevent the institution from luring back the consumers that new entrants to the market are already stealing way.

Illustrative Challenges / Opportunities

Cate-gory	Illustrative Challenges / Opportunities	Controls and Implications (The 4-V's) Volume/ Rate	Variety /Adapt	Velocity/ Freq.	Veracity/ Accuracy
1	Forge a connection using the fifth "P"—between FSBO personnel and the end customer.	↑	↑	↑	↑
1	Segment customers beyond traditional demographics and use targeted approaches aligned with delivery capabilities to go after micro markets that can be won and expanded upon.	↑	↑	↑	↑
1	Accept that brick and mortar approaches (waiting for the customer to arrive) are no longer viable methods to find desired and profitable customer segments / sub-segments.	↑	↑	↑	↑
1	Provide ancillary solution sets for research purposes, customer inquiries, mobile platforms, and any method the customer deems valuable to create efficacy and purchase.	O	↑	↑	↑

		Controls and Implications (The 4-V's)			
Category	**Illustrative Challenges / Opportunities**	**Volume/ Rate**	**Variety /Adapt**	**Velocity/ Freq.**	**Veracity/ Accuracy**
2	Create trust methods and measurements to promote loyalty and value-add across all messages and channels	O	↑	↑	↑
2	Reconnect traditional FSBO products and services to the channels and methods used by customers—only if applicable. Rework and repurposing are a given.	O	↑	↑	↑
2	Define and implement messaging that adds to credibility and market following rather than chase competition.	↑	↑	↑	O
3	Acceptance that many traditional marketing approaches are ineffective and new ones must be adopted due to how products are consumed, the regulatory compliance, and the underlying technologies.	↑	↑	O	↑

Category	Illustrative Challenges / Opportunities	Controls and Implications (The 4-V's)			
		Volume/ Rate	Variety /Adapt	Velocity/ Freq.	Veracity/ Accuracy
3	Minimal focus on fees and a majority of effort on value-add solutions sets (often for free or nominal value) in an effort to orchestrate a base.	O	↑	O	↑
3	CMO's need to embrace emerging roles beyond industry boundaries if they are to succeed.	↑	↑	O	O

↑ – Increasing non-linearly, unpredictable behaviors
O– Increasing but predictable

NOTES

"The bank gave us a 100 million credit line. Why won't you loan me a quarter?"

9

A New Financial Supply Chain

"When a government is dependent upon bankers for money, they and not the leaders of the government control the situation, since the hand that gives is above the hand that takes ... Money has no motherland; financiers are without patriotism and without decency; their sole object is gain." – **Napoleon Bonaparte.**

FSBO's are at a tipping point. Their bankers perform like a high-wire acrobat, tipping side-to-side, carrying a large pole (i.e., regulations) for balance to complete their tasks. The pole is also meant to spare their existence (to retain a safe and sound balance), all the while, walking across rocky crevasses on an increasingly thinning wire. Stumble and fall—you are no more.

Whereas, stress tests and guidance will hopefully protect the acrobatic banker from falling and killing someone in the crowd below (i.e., a safety net, if you will, protecting investors, consumers, and the bankers), the restrictive and myopic focus to force the finance individual to balance under immense loads will eventually destroy the highly skilled performer—or force her into retirement. After all, in today's world, financial institutions can only go forward or fall—there are few if any safety branches attached to their thinning wire path.

Nonetheless, why has the financial services industry yielded limited returns, while specialized segments (i.e.,

FinTech) have grown by thousands of percentages? Historically since 2006, the mortgage and finance industries lost well over 600,000 jobs and shuttered or merged 620 institutions resulting in a collective industry that is once again stable and generally out-of-the-headlines. Yet, this rebalancing has come at a cost—in capabilities, in flexibility, in opportunity, and in organic growth.

As a clear and present example, we can see this need for rebalancing in an area nearly a decade old—the government supervision of the GSE's and what to do with them in the future to maintain an affordable and fluid housing markets. Obvious questions surface from, "Are the GSE's ever going to be privatized?" to, "Is this something that the markets and future investors need to worry about?" to, "What structures will they follow and can the U.S. Government completely extract themselves from this complex conservatorship oversight role?"

Like others, we believe at some point the government will privatize those entities or spin components of them off. To this point, if you look at the various proposals from the Mortgage Bankers Association (MBA), to other groups who are pushing agendas, to the political base (who in theory should be protecting the taxpayer from risk and from future liability), there's a wide range of scenarios. We believe what's going to happen is they will be not necessarily privatized in the way that we think of private institutions who lack direct access to the Fed or Treasury funding.

In the past, the GSE's bundled individual loans and created various asset-backed instruments to ensure a balance between lending and investor demand. Unlike their prior structures and governance processes, future GSE-lite creations will be subject to a fair amount of oversight for very particular demographics by a federal government trying to guarantee access, lucidity, and transparency within a highly segmented housing market. It will be a new, but familiar, part of the financial supply chain but it will eliminate (hopefully) the GSE dominance across home asset-backed securities and the addiction fed lending institutions who don't want government involvement, yet also demand a method to close more loans.

If the U.S. Government finally allows capitalism to balance risk-assigned market demands (i.e., a good return for money to fund a loan), it will unleash a host of known (or forecasted) challenges and adjustments, especially for asset-backed lenders. However, if they are properly phased in against an iterative roadmap for change, a robust private market for lending will emerge without the speculation of prior boom-to-bust cycles. To create a new BAU within this segment of financial supply, FSBO's need to recognize that their position currently only promotes the reality of commoditization and non-value-added functions with a customer base that overwhelming views them with animosity. If all you do is take fees and push paper, then any FSBO can be made secondary with layered digitization and "upstarts" who bypass the need for control ingrained within many banking cultures.

Before you make any changes to the market, there must be an acceptance that the mortgage (as an example) itself is really a commodity—much like many of the traditional financial products that built today's FSBO brand names. Many products and services within FSBO's can be equated to household utilities—electric, water, sewer—they can be substituted, and most would be if homeowners are given the option (they are price sensitive). Regardless of the utility, they provide all the same basic things. The real issue is who do you choose as your provider and what do they offer to enhance, or redefine how the traditional financial supply chain behaves, interconnects and makes life simpler (and with full disclosure) to the consumer?

A Deeper Innovation: Where will Growth and Profitability Surface?

There are lessons to be learned from an industry that rival the totality of FSBO's in terms of investments and market presence. They have supply chains that span the world and they partner and outsource with those providing components to their solutions. It's the automotive industry, and they are now in the process of destroying their business models as a result of innovation and more importantly, because the consumer tells them.

You can look at thc recent pronouncements from countries such as Germany, Great Britain, and most recently China, all setting mandates that within 20 years, the combustion engine will be banned. What this says and does for automobile manufacturers, oil companies, parts makers, shipping companies, pipeline firms, and anything that goes into making, supporting and running a combustion engine, is your old models of operation and serving a customer will be obsolete. A few facts surrounding the alternatives—electric cars primarily and alternative fuel secondarily—will likely require 40% less labor, be easier to repair, boost electrical demand and cut perhaps 500,000 global jobs within 10 years at current production levels.

Even automobile manufacturers such as Daimler and Volkswagen have signed up to the radical destruction of their global business models. As the CEO of Daimler, Dieter Zetsche said, "In-house production is almost irrelevant to the consumer." One analyst, Richard Windsor stated the obvious for this industry consumer and environmental revolution, "Vehicle makers are queuing up to announce their commitment to electric vehicles, but at the same time they may be cheering for their own demise.[xciii]"

The parallels can be seen and conclusions drawn when it comes to traditional FSBO's and their decade of rebuilding since the Great Recession. While they have consolidated customers to a greater density then ever thought probable (i.e., the Big-5), the consumers are embracing rapidly FinTech solutions. The result is that those leaders burned in the fires of FSBO restructuring and regulations now face a changing market where they are angry at the very individuals (i.e., consumers) who support their margins. We can see this with JP Morgan & Chase Co. CEO going on a rant against cybercurrencies and then as a matter-of-fact, supporting blockchain sometime in the future (which many large banks hold unique patents secured since 2010). Perhaps it is a bit disingenuous to complain, at the very least, about advancements to secure additional time to compete?

So, as automobile industry segments now seem to be on a collision course to destroy their models and their finely-honed supply chains, FSBO's face a similar set of supply chain woes in the next few years. For FSBO's, they have recovered—like the automotive industry did post Great Recession. Yet, unlike the automotive industry, there has not been a single defining catalyst. However, for the FSBO's there have been a series of incremental catalysts which have materially changed their delivery solutions, their future offerings, and their profit models. FSBO's just have not linked all the dots together to create their "new picture".

The last three years have witnessed a return to consistent low delinquencies and defaults, however executive offices and industry groups cite risk and burden of regulation (e.g., Dodd-Frank) and consumer demand (e.g., lack of household formation) as factors in an anemic recovery of interest in financial services and products. Though the industry has shrunken in likely benefit of its reemergence[xciv], it has not, as a collection of linked industry segments (e.g., retail banking, custody, commercial, risk and security, trading and settlement, payments, technology, et al) been able to find cohesion due to long-tail restrictions and an overabundance of complexity that penalizes early adopters and frequently rewards mediocrity.

The very regulations that have (as some believe) made the financial supply chains "safer and sounder" have also created operating environments and cultures, which are unable to adapt and lead without explicit directions often for fear of retribution. Meantime, non-traditional players are working all sides of the delivery equation—consumers, technology, data, recovery, security, patents, software—as media coverage talks about how these innovators are finally deconstructing decades of decay in favor of the future. Maybe they are—maybe it is just another wrapper for a new act?

Who are the Real Customers in the New Supply Chains?

There are so many things we accept as fact when it comes to banking and finances, including how our products and services are built and consumed. As FSBO's, we are obsessed with the optics of what we offer, how our brand is perceived, where customers value our services, and even what regulators are privately saying to us. In truth, no one can fault bankers for worrying about optics given the negative media attention (much of it self-inflicted) during the last decade alone.

Traditional operating procedures necessitate that when potential consumers come seeking transactional products or asset-backed debt at our institutions (virtual or real delivery), we often want to know our customer (a broad interpretation of KYC) in detail. We ask them for information about their habits, existing obligations, ability to pay, personal identifiers, how they are using our solutions (or money), and anything else we need (or want) as part of regulatory demands and traditional risk management. For some consumers, bankers have become the Orwellian "Big Brother" (viewed as porous with data).

During the last decade, we've expanded our financial supply chain customer knowledge by asking or seeking out their social media and internet usage, competitive offerings, private information (often gained when they opt in to our apps down to their location and what they viewed), the type of phone being used, and even assembling third-party data to build a profile of every individual consumer. It does appear appropriate considering that global debt now exceeds all GDP by 325%—a total of $215 trillion bring forth elevated risks of default (see Institute of International Finance)[xcv].

If we're going to make changes to the financial supply chains, then within a free-floating capital market you need to minimize government involvement not just with regulations, but also acting as the primary intermediary for asset-backed instruments. If the government wants to retain a role within the markets, then a viable solution is to treat them like utilities. FSBO's in the future, given the disintermediation taking place up and down the supply chains, should be governed with broad principles like a public utilities commission. Looking at the market holistically, we should ask, "how do we guarantee that FSBO's are safe and sound as well as permit equal access and opportunity for those willing to live the consumer or American Dream?"

We've talked and talked about reforming the financial supply chain to benefit the customer within the last decade—believe it or not. It started in earnest in 2009 and with the bailouts (e.g., TARP) due to implicit guarantees, and we're coming up on a decade of GSE "utilities" under conservatorship. The problem will be is that after another decade of now explicitly guaranteeing loans, and the Fed paying a shell game in keeping trillions in MBS on their books, the unwinding of these assets with (now) implied utilities, however they're regulated, and governed makes the transitioning to a new supply chain of any type all the more risky and complex.

What has transpired is a complete fracturing of what capitalistic financial supply chains should be doing and regulators fixated on dogmatic responsibilities, and implicit politics. For consumers, all this talk reinforces a belief that they need to look elsewhere for financial solutions to simplify their life, to offer solutions how they consume them, and to integrate (not control) their traversing of the financial supply chains (as defined by the products and services they use) from retail purchases, to sharing of information, to guaranteeing their security and privacy, to offering them truly innovative financial opportunities (e.g., crowd sharing, ICO's).

And while the government highlights the need for "safeness and soundness" across the financial supply chains, opponents to change (and those benefiting from inaction) highlight the "unknowns", the risks, and unregulated markets (which many want freed from regulation, but don't want any competition). Proponents for new financial supply chains, cite consumer sentiments, product acceptance, cross market successes, and facts that they know their customer better than most banks (it's more than just fee-based income to them). It's a set of big unknowns right now because talk rules rather than action. Those concerns are going to take a long time to understand their actual risk impacts on the balance sheets—we can model and model, but consumers will not likely wait.

A Deeper Analysis: Questioning The "Facts"

Are the headlines the realities? From cross-border issues, to valuations, to business models, financial services have never been as much in flux. Institutions are undergoing reimagining and rebuilding into structures that are both fluid and rigid (i.e., fluid to meet consumers and rigid to satisfy regulations). If we broke an institution up along 12 distinct but generic continuums, we would not only gain insight into markets being served and likely profitability, but adaptability, and even capabilities. A representative set might include:

- Layers of needs
- Layers of solutions
- Layers of regulation
- Consumer shifts
- Data and digitization
- FinTech
- Market sentiments
- Economic uncertainties
- Privacy, security and recoverability
- Analytics and measurements
- Availability of skills
- Legacy, legacy, legacy

Moreover, by further decomposing an organization along lines of business (e.g., mortgage, asset management, brokerage and security), another layer of insight would be gleaned from the blackboxes containing the institution's inner workings. Now add another layer of experts, advisors, and political pundits and you have a delivery mix that quickly becomes murky and prone to long-lead times to "get it right".

FSBO's profits are still very positive, so why hasn't the workforce and level of pay rebounded? What is increasingly bipolar is that institutions are booking profits, while at the same time reshuffling workforces and shedding more jobs. Oliver Wyman stated in their annual State of Financial Services Report that, "Average returns of large financial firms have fallen from over 20% in the early 2000s to 7% in 2013, the level of utilities companies." Even as productivity per worker stalls, the idea of doing more with less likely because survival scars are having an impact on organic growth potential as consumers shift their usage and behaviors.

Why Innovations are Integrated into New Iterations

Why have financial institutions created skunk works, formed internal innovation teams and funded venture capital initiatives for technology advancements? Since 2008, headlines have increasingly built on the theme that financial technology (FinTech) firms are the wave of the future. This is true for many, as yearly FinTech investments went from under $100 million to over $50 billion globally—a five-hundred-fold increase in over a decade. For FSBO workers, it has been the ultimate insult. They survived the greatest threat to their existence since the Great Depression only to watch investment money and their own executives shunning internal growth in favor of lift-and-shift apps and solution sets. Fundamentally, it has become the new global outsource process—without the people—due to technology advancements in provisioning (e.g., cloud), security (e.g., ported to specialty vendors), and app and API vendors (i.e., those who provide the interfaces and glue along with consumer contact landing zones). Moreover, with automation rising, their jobs are increasingly at risk[xcvi].

Can anyone blame the executives and even the speculators seeking out the next growth curve across financial services? As institutions hit with layer after layer of regulation coupled with multi-year litigation and government fines, they are afraid to take risks that do not come with an explicit "get out of jail free" card sanctioned by a regulator in writing. A quick example of this is the global negative brush of payday loans, which has created a shadow banking system for those deemed unacceptable to society's core values (talk about unintended consequences for the approximately 80 million unbanked and underbanked consumers).

So, how should we look at FSBO's? Are FSBO's truly commoditized? It is no different, if you take that line of thought, and say that in the future banking is a series of platforms or a whitelabeled set of services, which can be wrapped with innovative, differentiated solutions already familiar to consumers. If you begin to examine where FinTech firms are gaining traction, it concentrates around areas of financial supply chain efficiencies, automation, and superior customer interfaces and product capabilities.

With the changes in politics and perceptions, even the large outsourcers like Infosys, Wipro, and Cognizant are indicating that banking as a utility may hold the future of financial supply chain capabilities when considering the large infusion of non-traditional FSBO offerings and consumer preferences. When we examine the politics of hiring domestically and getting the American worker more involved with economic prosperity, we believe this lends increasing support to the iterations of this utility concept where you use common infrastructures, common processes, common depository type of instruments, and tightly-coupled applications.

There is a precedence from industry groups who have created standardized interfaces, API's and data repositories (e.g., MBA) to facilitates data interchange, data transparency, data due diligence, and all the forms or "wrappers" that encapsulate their outputs. These collaborative advancements allow FSBO's and non-financial entities to enter into the market, understand the risk, and create new evolving instruments in the lending market that will be post the GSE's, and part of a revamped supply chain.

We believe that there are going to be a series of events, probably not linear, but a series of changes taking place all along the financial supply chains including how changes or capabilities are provisioned as upstream and downstream impacts will be felt and required.

If you're applying the utility model as a part of solving government in asset-backed securitization[xcvii] to deal with GSE disposition, you've to realize that this fundamentally changes the banking model itself and how you look at asset valuations, assign risks, offer products, and deal with innovative destruction of traditional supply chains. Moreover, this implies that we accept natural commoditization of offerings such as once-prized mortgage lending and car loans.

What makes the difference? That's where you basically recognize that financial supply chains are required to service a base need much like a computer, it's plug and play. It allows the advancing consumer to ask, "Who's got best or more innovative offering beyond base functions?" For FSBO's it forces the questions of, "How do I partner? What are those partnerships? What don't I have? What do I use internally? What do I do with partners? What do I do externally to create a holistic service concept when we don't control the entire supply chain or channel?"

That also launches the next questions, which is implicit in some of our prior questions, "Is it all about owning the customer? Or, is it all about serving the customer?" Those are some big distinctions if we use the secondary market as a common example, which are going to permanently change FSBO offerings and more importantly, their behaviors. As we have indicated, these transformations are not going to be an overnight change. The rationale for not transforming the entire chain is a result of several truisms;

1) the entire chain is amorphous moving forward depending upon how and where you look,
2) selective transitions allow for iterative improvements that can be compartmentalized across increasingly complex delivery,
3) organizational cultures cannot adjust or internalize to divergent digitization demands, and
4) if you get it wrong and the markets themselves begin to tank because of transformation momentum[xcviii], it becomes another form of government's control to come in and bail out (because if the financial supply chain freezes, money lending, and transferring of balance sheet items to an individual investor, then you've got some serious issues. Basically, nothing flows and the result could be worse).

A Deeper Reality: Accept the Limits and Move with the Pressures

For the supply chains implicitly embedded within FSBO operations by the processes, directives, and technologies selected, the delivery chaos and organizational resistance can be severe. The realities facing FSBO enterprises brought on by capability limits and pressures are no longer about an organization adapting to one or two hazards to their operating models—there are many threats at once.

To put this complexity into perspective, let's look at an illustrative graphic applicable to most FSBO's, and their asset-backed segments (e.g., mortgage auto, boat).

illustrative

Limits / Pressure Forces

Time-to-Market
Data Archiving and Retirement
Structured Strategies
Pervasive "Model" Iterations
Time-to-Value
Customer Loyalty
Change Resistance
Managing the Optics
Decision Making
Assessment Accuracy
Governance and Oversight
Investment Capability
Customer Behaviors / Churn
Principles to Granularity
Timing to Returns / Profitability

New Financial Realities

	Old	Transition	Transformation
Revenue Goals	Maintain / Retrench	Cost-Efficiencies/ Cost-Reductions	Multi-faceted Layering/ Relationships
Operating Channels	Singular	Mobile	Omnichannel
Deployment Models	Shared Services	Utility	BaaS / BaaP
IT Reusability	Build	Expand	Consume
Services	Best-Practices	Collaborative	Value / Customer Defined
Standards	Definition	Layers / Interoperability	Complex Layers / API's
Compliance	Risks / Process	Attribution / Reporting	Adaptability / Transparency
Partnerships	Cost of Operations	Mitigation of Risks	Quality / Service Capabilities
Technology	Application Driven	Mobility Enabled	Omnichannel
Data	Consistency / Standardization	Compartmentalized / Data isolated	Orchestrated / Layered
People	Skills	Team / Individual Contribution	Innovators / Adapters
Organization	Owned / Hierarchy	Function / Business Area	Virtual / Dynamic and Variable
Digitization	Fundamental	Complex / Discrete	Complex Layering
Customer	Ownership	Minimally Shared	Custodian
Processes	Automation / Core	Value-Added	Value-Efficiency / Open Standard Assembly
Performance Gauges	Metrics / Dashboards	Trends / Forecasting	Patterns / Anticipation
Security	Add-on	Overlaid	Tightly Integrated / Loosely Assembled
Intelligence	Limited / Silo	Predictive	Machine / AI / AR / VR
Privacy	Policy Based	Corporate Protection	Consumer Controlled
Supply Chains	Primitive / Fragmented	Automate Existing	Focus on Value-Add
Marketing	Point-to-Channel	Channel-to-Channel	Any Channel

Note: There is an implied one-to-many processes per supply chain. Therefore, a supply chain can incorporate many processes.

What is immediately apparent, whether you read every entry or not, is that the complexity of FSBO change is well beyond the ingrained ideals of transitioning an enterprise, which cannot be accomplished by a one-off change or acceptance of channel or efficiency goal. Today, FSBO's face a fundamental and comprehensive redesign of their own supply chains starting with the customer and working back into every process and procedure.

Customers Demand Collaborative and Diverse Supply Chains

It is this approach of moving through the supply and delivery chain, assessing existing to future capabilities (i.e., existing, build, partner, acquire), which will provide the greatest value not just for efficiencies, but a clear understanding of "can we compete?" If the answer is yes, then the "how" booking positive results coupled with rapidly transitional of "what" we will provide, creates a map for expansion or redevelopment. An example of complex data product improvement designed to delight the consumer, as McKinsey noted in their digitization of business process brief, is where a bank case study cut the "cost of a new mortgage by 70% and slashing time to preliminary approval from several days to just one minute." Reworking and augmenting the entire supply chain was mandatory.

As we iteratively conduct assessments, taking to account the large number of variables and internal politics, the reality of financial actualities appears overwhelming—just too hard to do. Beyond the bravado of "we will dominate the market..." the layers of components from operational intelligence, to strategies and patterns, to regulatory compliance, to risk analysis must be matched to partnerships, JV's and business delivery capabilities, and even our maturity with complex, layered digitization.

And then, within each of those segments they are further decomposed into additional functional categories, which demand further scrutiny before a single line of code can be cast. It is precisely why IT during the last two decades has become so fragmented within and across business lines, and why vast amounts of time and capital are lost to "white elephants" reminiscent of custom building the old TDA and DDA systems decades prior. We run for the trap of technology without seeking the relevance, and more importantly, how it impacts the supply chains and business models.

However, the collaborative realities of tomorrow raises some more questions in our minds. The first one, in relation to what we said about owning or serving, is it realistic anymore that banks will be able to "own" customers? We know they've traditionally had and perceived it as a given, until the rise of non-traditional competitors. FSBO's continue to talk about it, strategize for it, and allocate precious resources towards it. They talk about share of customer's wallet, but in a commoditized financial services world, what's the chance that they're going to be able to have that as a defensible position?

Our initial response would be that we don't think they've owned the customer for a better part of 15 years—at least. But, they haven't quite come to that realization, or more than likely, they haven't quite come to that realization yet on their own. To them, and from how they generate fees off every transaction, and in some cases interactions, how FSBO's service those customers immersed in a virtual or numeral world with digital currencies will determine their profits as a value-added institution along the financial supply chains. As an example, the Bank of England is working on a virtual currency, which will disintermediate or eliminate some of the money management functions found with paper-based banking transactions. If reduced or eliminated, you begin to look at these discrete actions it begs the question, "Who really owns that customer now and in the future?" If paper money and physical checks (the latter representing < 10% of all transactions), continue to be shunned in favor of their digital equivalents, then the entire process supply chain becomes disintermediated and replaced with solutions outside FSBO control.

Additionally, the customer doesn't necessarily value traditional, closed loop relationships where they can only select from a limited ecosystem of financial supply chain offerings. This need for control by FSBO's was reinforced from the early days of cross-selling, and as we noted, a share or command of a consumer's "wallet" (e.g., credit cards, auto loans, home loans, boat loans, HELOC's, college loans, investment management, stock trading, et al). Even within corporate goals linked to the FSBO's go-to-market strategy, were ideals that said, "the more I get from that customer, the better off I am. I can lock them in and reap relationship fees and income for the entire duration of their banking activities."

If you go in with a "lock them in" mentality, that says we're going to own the customers, most FSBO's will be dissatisfied with any partnering initiative, and will likely fall short of desired market and profit projections. In most FSBO's go-to-market efforts, which we talked about in the last chapter, develop and deliver their apps (sometimes explicitly) with a singular theme of, "I'm going to control the customer's financial offerings. I'm going to lock them in.

I'm going to upsell to them and discourage through penalty or restrictive offer anytime they don't follow our rules." And we know as customers and FSBO's, it takes consumers a lot of effort to move from one financial enterprise to another. If FSBO's continue to pursue control features, they will alienate the customer in across today's diverse markets because financial consumers are far savvier and less forgiving than previous, aging, and high-touch demographics were.

If we, as FSBO's, use our brand, delivery systems and fine print to restrict financial consumers, or prohibit them from utilizing emerging financial products (i.e., ICO's as an extreme example), we will find our prospects dim when compared to other industry participants who place the customer first across the various supply chains. Any institution that is being a "restrictive parent", trying to lock consumers in, to make sure they can't transfer or get fees for transference, will pay a price not just in the media (e.g., social P2P headlines) but as being part of the institutions who are commoditized and marginalized by the very demographics they sought after.

As financial consumers look at their options and merge viable solutions with their lifestyles, especially in this post-secondary and broadly integrated supply chain world, FSBO enterprises that not only meet app preferences, but those who respect consumer business no matter how large or small, will be the institutions who in the end have delivered greater supply chain capabilities.

An indicator of why we believe the financial supply chain will favor those who embrace collaboration centered on the consumer there is at last count over 15,000 different APIs that one vendor alone has compiled within their integration repository—a lot of supply chain options.

A Deeper Demand: Road Mapping the Future using BaaS

Yet, when our enterprise maps desired changes to meet consumer demands, additional levels of necessities become apparent. The following graphic illustrates the sophistication of analysis necessary for supply chain examination, and how some of these categories interact (e.g., plug-and-play) with other organizational demands.

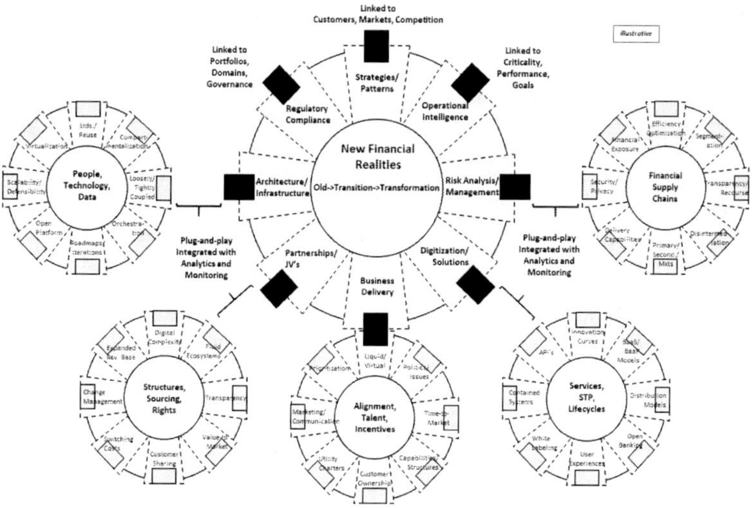

Some have written that this redefinition of internal supply chains is a natural evolution as enterprises apply simple digitization and move into multifaceted, layered digitalization. Whatever you call it, it will be the future for FSBO's. For FSBO's trying to negotiate BPO or ITO contracts in the middle of transformations, Cognizant's "How Banking as a Service Will Keep Banks Digitally Relevant and Growing", offers a unique discussion from an outsourcers viewpoint addressing some of the plug-and-play items.

Big is Not Always Beautiful or a Recipe for Supply Chain Success

When you look the number of API's and the discrete elements they represent, you begin to imagine the size and complexity of the data repository. Can you imagine how information flows and its potential and varied uses? Its analogy is the ability to plug into an outlet, get encapsulated set of information that is always consistent, always with the right expertise, and able to deliver greater FSBO products and services assembled from stacks of solutions (think of a refrigerator, a microwave, washing machine, game consoles, AI or voice units for the home). It's "electricity", a commodity, but we can now transfer our electricity to a different provider—even though the transmission lines, the back-end pipes of the front office, the middle office, and the back office of traditional banking are all the same. How we go-to-market to reach that financial consumer, and how we provision that for a customer, that's the difference of how we view that customer. Are we going to lock them in? Are we going to provide them value, excellent value, value-added services across our components of the financial supply chains?

If you look at it from two different principles or perspectives, we think that's a fundamental problem and on-going misconception in this post-secondary, post-GSE world. FSBO's are struggling to find their role in the future given that they are profitable today. But we think the lack of a leadership role is a compelling causality of why a majority of FinTech's have gotten so much interest and investment. Think about it. If we review the $25 billion in the last year alone that was put in FinTech startups, add in all the things they're trying to accomplish—from obtaining banking charters, and from the OCC trying to regulate their purpose—the problem we get into is the more you try to regulate them, the more ways small innovations will have a huge impact on the "Big-5" as consumer innovation can only be held in check for a short time.

FinTech's are looking at financial tradition and with commoditization and disintermediation, recognize that the financial supply chains are not designed for a globally connected consumer. There is a fundamental shift or fracturing taking place here. If we're expanding technological consumer solutions, the expectations and perceptions of those consumers is that FSBO's will step up and meet their needs wherever they are regardless of physical location. We've vast numbers of API's from within and across industries. We've got data standardization spurred by industry groups and associations. We've got information warehouses that swell 24 by 7 and move from siloed datasets to epic, big data solutions that aid machine learning and AI. Reviewed comprehensively, being a big FSBO is really a liability with the markets and from the perception of a growing number of financial consumers as these industry dominators are viewed as resisting change and restricting their demands.

"Big is beautiful" has been incorporated into a series of corporate strategies developed and refined going back into the 1970's. Just because you are large today the ability to "sole-source" the customer's usage is waning, you can't control their appetite, you can't wait out the markets, and you can't internalize a false belief where "brand trumps" the reality of the offering. Today, FSBO's can provide a tremendous amount of transparency, we can partner with greater success if we have the assets of a large institution, we can offer diverse and bespoke solutions, and we can have empathy that we, and the consumer, are in a very fluid financial supply chain set of offerings. We can make the changes needed for supply chain retooling if we take the fundamental shifts seriously—just as the auto industry is doing with electric cars.

However, that also implies that we are consistently delivering (and adapting) high levels of security, high levels of privacy protection which needs to be embedded in these transferences, and higher due diligence when sharing precious consumer financial information with those connected to our trusted and layered networks. We have also witnessed the consumer having a progressive appetite for iPhone's, mobile banking, and a dramatic decrease in "paper financial products".

If we examine the cultures and principles at the core of most FinTech's, these new startups are laser focused on consumer needs, usage, and perceptions. They don't believe they will fail, and the way they compete is to circumvent the establishment—much like Bernie Sanders did during the presidential primaries of 2016. They don't believe they will fail and their followings have become cult-like in their continued enthusiasm for greater capabilities and broader delivery options.

These changes usher in new realities for fragmented, large institutions or those historically built by a hodge-podge of M&A layered applications and back office systems. How many times have we (FSBO's) already said we're going to add a security package? Or, we're going to add a privacy oversight committee, or set of APIs'? The future is not about defining and plugging these interfaces and exchanges in after the fact, and often sloppily (perhaps this is why one of our identities were stolen and no financial institution even notified us?).

The future represents those emerging (and "stubbed out") features and functions, which are plugged in from the start. This new series of approaches, starting from concept to design to delivery, is a big difference and really what we know will be key differentiators of how FSBO's operate in the future.

From commoditization, we really are getting to that future world based on technology, and based on consumer habits of how they use financial services. The idea of clinging to a brand over substance is like holding two pair against a royal flush.

We think those are the big changes that FSBO's (large and regional) are going to have to accept—and begin to embrace. These differences, be they platforms, be they services, be they combinations of currencies and settlements, are forever changing the traditional supply chains—and the consumer doesn't care if you have provisioned for these changes, they just know they will take their business to those institutions who offer the solutions sets. We already know that a FSBO doesn't necessarily need to be a physical institution anymore (with brick and mortar fronts) either, so if the idea of what constitutes a financial institution or bank is virtual, don't you think the supply chains will also become less traditional?

A Deeper Problem: The Entry of Larger Cross-Industry Competition

As a rule, we've greatly increased knowledge about our customer "supply chains"—but we are surprised that these same customers increasingly view decades of back-office processing or even our entire institution as a "utility" or service that adds minimal value. As FinTech advancements push front and middle office solutions into successive iterations driven by pervasive mobility and robust data intelligence, consumer relationships with banks themselves has become one of commoditization—"anyone will do if it fits into my apps." Stated as a consumer might evaluate us (FSBO's), "What unique value do you provide?"

This creates a series of problems for FSBO's. In the past, the challenges to their internal fulfillment or achieving supply chain efficiencies, were between traditional competitors under the watch of regulators. Now, consumers are using P2P exchange and settlement solutions, and fixating on a mobile platform rather than the iconic capabilities FSBO's built over decades. Factor in the tens of billions put into startups in the last three years designed to replace vast portions of our infrastructure, while shifting consumer touchpoints to third-parties, and we seemed surprised that over one-third of customers would rather bank with a retailer or social media giant—than a bank itself.

Finally, it might be heresy to some, but the introduction of regulations may have slowed transitional needs to achieve market-driven transformations (as noted in the first diagram above). Moving forward, if the regulations are relaxed or not-enforced, then the competitive pressures alone over the next 2 to 5 years will force many institutions into shot-gun M&A's or to redefine their business models as utilities allowing others to deal with value-added and highly fluid customer interactions. In fact, an argument can be made that the total number of banks will continue to shrink year-to-year even with the recent application of new banking charters, due to a failure to comprehend necessary material transformations needed within and across financial supply chains.

Moreover, FSBO cultures and reward structures are driven towards "high-producers" or those who can increase the top line. The enterprise models are constructed around individual customers and how we acquire, cross-sell, and up-sell to them—"a one-stop shop". Yet, the model emerging is one where FSBO's are not, and likely not to be, the starting point for customer contact. The future FSBO models are increasingly based upon layered collaboration (not just top performers), consumption not construction, and information standardization with a virtual staffing model.

BaaS (Banking as a Service) Requires Learning a Few Lessons

Banks are fast becoming just one more service across morphing financial supply chains within a global network of delivery components that can be accessed via API's, app stores or even game consoles and virtual reality. As the multitude of offerings increase, it's going to make FSBO's learn some emerging lessons regarding expertise with aggregation, compartmentalization, and servicers, which banks have been using to perform discrete, non-value-added duties.

We envision that there are unique lessons from these inter and intra supply chain partners that FSBO's can start to apply to practices from not only how they perceive their place in the financial world order, but also as transformation roadmaps, to a banking-as-a-service (BaaS) model for commoditized items, and those that can be used as part of orchestrated stacks for new products and services.

From our perspective, FSBO's need to ask themselves, are they willing to accept the lessons coming from the deployments with BaaS? It's like anything else. If we do not accept the realities of FSBO supply chain destruction (coming as part of a BaaS approach), if we do not internalize what added-value we offer, what then will be part of the competitive action plans? If we cannot put into context the changes being made not just by us, but including our traditional and atypical competitors, the "advancements" we implement may just represent an exercise of how not to adapt to market dynamics and ecosystems.

If we undertake a framework that fails to first address baseline success requirements (e.g., like a farmer preparing a seedbed), we try to be everything to all financial customers, and we try to control relationships like we've done in the past, we're not going to be very successful because those organizations or entities or leaders who are more flexible and embrace the environment will thrive. BaaS for this discussion, represents a significant departure of how systems are not only provisioned, but for how we embrace collaboration, go-to-market and even and most importantly interact with our financial customers along the supply chain.

Moreover, how do you deal a series or products and services which are offered as part of a virtual bank? How do you offer different liquidity solutions to retail and commercial customers? Do we accept the open standards of banking and platforms? Do we get rid of the belief or dispel the accepted fact that digitization by itself, is not an end all, one-off solution? These are just a few of the paths to get us to new operational iterations, but they require foundational elements and a clear set of experiential learning, which can apply BaaS methods and techniques to carve-out front, middle, and back-office solutions continually being reworked by shifts in technology.

For instance, when we look at our operational attitudes and capabilities from a lending standpoint, where is most of the investment money going to come from (to create whole loans in a portfolio)? The trends are increasingly showing that the vast use and deployment of BaaS is going to come from offshore. In using all types of lending products and their asset-backed cousins, accepting that traditional lending is being commoditized, and accepting lessons from (traditional or non-traditional) competitors, FSBO's should realize that this future is not the world we were just contemplating even three years ago. If that's the case, then how are we going to compete in the future if we don't alter our delivery methods?

As FSBO's we don't want to say we've been blessed that we've had regulations. That's a weird, not wonderful, thing to say. However, regulations have caused a huge advancement with understanding what is going on in our financial supply chains, how we track and report assets, due diligence mandates, and many more critical functions. What kind of data do we have? How available is it? What is its quality inside and outside our networks? How auditable is it and what legal liabilities linger within its explicit or implicit existence? If we never experienced regulations and the ugly umbrella of Dodd-Frank, would FSBO's ever have gotten profitable, and even able to address what markets and partners they need to compete across financial supply chains?

Now, if you roll some of those rules back, or you begin to say, well, "do I need them", it positions those innovative FSBO's or FinTech's with a competitive advantage[xcix] especially in the face of political flip-flopping surrounding regulatory mandates. These innovative few institutions believe, "We can short circuit these regulatory compliance needs, and you, FSBO, do not have to make the same mistakes some of these other people did as you exceed the letter of the law and meet its spirit of intent."

BaaS starts with some fundamentals when viewed as a catalyst to aid with material financial supply chain improvements (i.e., varied, simpler, multi-faceted, international, tailored, adaptable, transparent, standardized, and dynamic—did we miss any?). To begin, we should distinctly define, "What are we in the business to do, and what types of models do we have to make that work?" You should not view this delivery as a consulting exercise or a book prescription model using hierarchical or matrixed decompositions complete with an endless supply of meetings, focus groups, and anything else that can chew up months of elapsed time—you don't have months to "think about it".

Moreover, there is the realization when FSBO's dig their heels in, taking their time, not being nimble and open to suggestions, is when they fail. When FSBO's embrace change brought forth by BaaS and utility models (along with their operational processes implicit within the supply chains), they value and understand standardizations, they understand consumers reactions, they understand the legal challenges and regulatory mandates, they understand the political chaff and rise above the concerns. When FSBO's put these challenges and opportunities into a framework, will they become conductors—as part of a capital-lite orchestration model.

FSBO's traditionally approach supply chain transformations from the top down. When you look at it from the foundational elements which are transforming below us, what transpires is that the top-end model (that is highly valued), doesn't work very efficiently. We think that's where bankers, who are such a conservative bunch, must begin to comprehend that they are in an on-going commoditization of the financial world, especially in areas such as lending, on-demand accounts, and even investment management. How these elements align and emerge in what FSBO's offer is going to be a huge difference across the financial supply chains moving forward.

A Deeper Mandate: Deliver Across Many Channels

Whatever the circumstances facing your FSBO, and before we chase solutions, we should examine and integrate them into our own complex supply chains—especially if they fit without the supply chain being materially reworked. For FSBO's to remain relevant, the traditional organizational approaches to achieving operating efficiencies (i.e., not touching the internal supply chains), lack efficacy as BaaS and utility models disintermediate and commoditize cash-cow offerings while alienating financial customers.

Additionally, if, and when Dodd-Frank and the hundreds of individual mandates are removed or changed, those FSBO operations longing for a return of the glory days, may find themselves marginalized. To assume there is a value in rollbacks, there is an implication which includes a steady-state across their operating environments and customer interactions. As we have seen in the preceding diagrams and chapters, none of these plug-and-play categories have remained constant, and the removal of regulations will likely yield a vast amount of expensive chaos—and perhaps a call for re-regulations.

A method to avoid chaos, is to transform internal supply chains with a business model that provide utility processing and BaaS features, while concentrating on consumer value-added solutions. The "ancient ways" of banking are gone. Reinstating them with regulation is no better than with regulation. Consumers are calling for new omni-channel models, new supply chains—are we prepared to positively answer them?

Demanding a Blending of Cultures, Technologies, and Channels

So where are we now with the rebalancing or redefinition of existing financial supply chains? Do we pretty much have the technology we need to deal with all this now? Or, will there be new solution sets helping FSBO's in the future?

We believe right now there is currently a wide variation of hardware, software, processes, and even AI in the markets today—enough to deal with current requirements, while leaving build-out stubs for future capabilities. There is a vast amount of options for FSBO's in terms of how things are changing and shifting when it comes to supply chains, and how they are provisions, modified, and disposed. Dealing with automation and virtual reality, the practice of plug and play encapsulated solution sets, and even finding the best fit for APIs, has created a deficiency within internal staffs who are overwhelmed with choices and implementation complexities without strong partners and pragmatic advisors.

We ascertain that most of the solutions sets (encapsulated technologies) today exist to support various forms of BaaS and utility approaches as required to operate in a collaborative FSBO enterprise. Will there be new things that are developed, which will alter today's "knowns"? Of course, because the way we deal with information, and how we deal with it, and who we share it with, how we share it, how we securitize it, and the security of that information itself, continues to advance spurred by a financial consumer who is anything but traditional. It's going to be a evolving, iterative process where short-term, transitory planning will be a greater asset than "thinking BIG". That is something, again, that is typically counter to the traditional culture of banks. FSBO's are used to building or buying an application solution, holding it for 5, 10, 15, or 20 years. As we know, that doesn't work today or tomorrow—and likely has been an anchor for years.

We're referring to interfaces and delivery systems which that are provisioned nowadays that may only last six months, a year. We have the technology and we need to be rapid-cycling it to meet financial supply chain mandates. We have the capability to make sure that data and compliance information is available even if the product or the service has been retired. That is a whole different concept to being nimble, quick, active, and having due diligence within FSBO environments, which historically were culturally conditioned to be "producers" concentrated around coveted "rainmakers".

Hence, the idea of having architectural constructs linked to dynamically changing infrastructures were secondary to the business unit head who believed, "I'm going to build everything and then the consumers will flock to 'it'". However, while that worked until about 2014, today's definition and delivery model shows that as FSBO's we're going to begin to layer BaaS and utility offerings to achieve transformation and transition on a constant, recurring basis—regardless of existing products and services. That is a different formulation. This is the future of financial supply chains.

And what are these new financial supply chain investments chasing? The three obvious answers are efficiencies, growth, and profit (by layering attributed risks and isolating contamination). Additionally, strategy models and M&A ideals still fail to properly account for hidden complexities within legacy worlds especially when integrated with emerging API's, compartmentalized FinTech software, and next-generational experts. Bankers for their part, using familiar themes often prefixed with "e-" or "i", continue to amass solutions that while innovatively developed, are often obsolete or missing the mark upon delivery. Constrained by conservative organizations desiring to remain the same, how many bankers are willing to stake their careers on investments that are potentially written-off within their first year of their delivery?

Banners of innovation, fast moving themes of innovation (e.g., cloud, digitization, big data, AI, VR), and even partnerships with technology brands, may create headlines, but unlikely will allow enterprises to transform themselves from a slowly developing financial utility to a fast-adapting financial innovator across widely diverse supply chain requirements.

In summation, we are embarking on a new model of financial services within five years. Some of these will involve traditional bank models, which are primarily utilities—and other institutions that are technology providers who feed upstream and downstream parts of the emerging financial supply chains. The utilities will be the plumbing or back-office delivery behemoth controlled by regulators under the auspices of "safeness and soundness." This divergence should be recognized for what it really is—regulation is about managing utilities—regulation is almost never about true innovation (as regulation comes after "someone gets killed...").

Those who stay within and invest among FSBO's will be forced into decisions of "now that they say we are grown up and regulated, is this what we want to be?" Is this how we should serve our customers? Is this how the customers value us across the diverse financial supply chains? Are we prepared to learn from others so that we can leverage the existing, while avoiding the traps?

Illustrative Challenges / Opportunities

		Controls and Implications (The 4-V's)			
Category	**Illustrative Challenges / Opportunities**	**Volume/ Rate**	**Variety /Adapt**	**Velocity/ Freq.**	**Veracity/ Accuracy**
1	Developing adaptive capabilities to understand and model various FSBO supply chains ensuring that iterative improvements and rapid retirements can be achieved as part of their designs.	↑	↑	↑	↑
1	Creating compartmentalized segments with supply chains to target / deliver improvements in addition to being able to offer layered communication between encapsulated models and processes.	↑	↑	↑	↑
1	Developing partnerships with customer facing apps to continue relevancy and channel delivery—forgoing a dogmatic mandate to completely control the channel (and where expertise requires licensing).	↑	↑	↑	↑

Cate-gory	Illustrative Challenges / Opportunities	Controls and Implications (The 4-V's) Volume/ Rate	Variety /Adapt	Velocity/ Freq.	Veracity/ Accuracy
1	Using digitalization, not just digitization, to improve reach and profitability as rudimentary methods evolve into complex layering across channels, apps, and API's.	O	↑	↑	↑
2	Map out new financial realities to influence strategy, architectures and actions across all supply chain segments while promoting standardized solutions differentiated by value-added capabilities.	↑	↑	↑	O
2	Creating reusable, orchestrated app components that are compartmentalized, cross platform, and able to be layered across supply chains.	O	↑	↑	↑
2	Avoid "lock-in" methods and mentalities using BaaS and BaaP principles across delivery chains to gain competitive advantage and first-mover capabilities.	O	↑	↑	↑

Cate-gory	Illustrative Challenges / Opportunities	Controls and Implications (The 4-V's) Volume/ Rate	Variety /Adapt	Velocity/ Freq.	Veracity/ Accuracy
3	Creating reward and employee structures beyond being high-volume producers provides cultural internalization of orchestration methods (along with the building blocks needed for compartmentalized, and constantly changing financial supply chains).	↑	↑	O	O
3	Deliver layered collaborations with third-party providers allowing for expanded bases of operations and customer contact beyond those invented with the FSBO.	O	↑	O	↑
3	Align delivery capability for supply chain assembly with changing consumer sentiments and anticipated provisioning demands.	↑	↑	O	O

↑ – Increasing non-linearly, unpredictable behaviors
O– Increasing but predictable

NOTES

"I'm naming you VP of Revolution, Action and Edgy Thinking ...
on one condition ...
that you promise not to change anything."

10 What's Next?

"... The modern theory of the perpetuation of debt has drenched the earth with blood, and crushed its inhabitants under burdens ever accumulating." – **Thomas Jefferson**

The last decade has produced countless innovations across FSBO's—VRU's, ATM's, due diligence capabilities, kiosks, biometrics, standards, mobile banking, loan processing, funds transfer and many more technology enabled solutions within payments, deposits, settlement, customer service and credit processing. In fact, nearly every idea being pushed to executives has "innovation" somewhere in their title. Innovation is the answer to all the problems—yet, it has become meaningless in its overuse—"Innovation to grow organically ...", "Profit by innovation ...", and our favorite, "Innovation will expand our customer base ...". Seldom is the question asked, "What, why, how and does it make sense to adopt or create an innovation—will it help?"

Moreover, if innovation is clichéd, then the idea of digitization is also worn-out as most banking innovations, especially when factoring in FinTech announcements and media attention, involve some form of big data digitization. It seems the future of banking is all about automation (goes back into the 1960's), efficiencies (going back into the 1980's), redesigns (starting in the 1990's), and mobility (now going back around a decade). Nevertheless, is the future of banking singularly about one of them, several of them or is it something new based on the culmination of history?

As we've covered in the nine previous chapters, the future of FSBO's resides with recognizing that a framework for transformation is required. But it's not a framework that comes with volumes of explanatory sections, it's not a framework lifted from a competitor, and it's not a framework that takes armies of staff to achieve. It is a framework to gain understanding of what products, services, and customers add to growth and contribute towards profits. It sets filters to analyze ideas and innovational advancements. It is where you're at today, what you want to do, your current state, your future state, and a starting line for a roadmap.

Oftentimes we are so hung up on the lure of innovation, now wrap in the buzzword FinTech, that we really don't ever concentrate on anything that is required internally to make the innovation grow and adapt. We focus on the competitors, we focus on the FinTech, and we assume and wave our hands around and say, "All this 'stuff' inside gets fixed, too."

The problem is, that's right now many FSBO's allocate between 85% and 95% of operating budgets on BAU, and most of these new program initiatives are gone within two to three years. They have either failed or they've been retired. So, the rapid cycling of them, if you will, is much greater than it was in the past where we could create a check-processing system or demand deposit system (i.e., old school), and they'd be around for 20 to 30 years. It doesn't really work that way today, but the cultures propped up by legacy ideals foster a false reality. By taking the idea of this framework (further presented in detail in diagrams later in this chapter), we begin to say, "What do we really need to do? What are the priorities? What are the limiting factors? What real benefits are we going to get from taking this calculated program risk?"

In every organization, FSBO's can tailor our framework to become their own, and they can expand it or contract it as they need. The framework of how you approach these changes, filter these innovations through, works for advancements such as Agile deployment very well, which a lot of organizations have sought out to improve product capabilities, cycle-time, and quality. The use of a change framework moves beyond the euphemistic “magic happens here” button of, "Well, it's innovative" or "Well, we partnered with somebody, and that's how we delivered our front-end processing application for our mobile customers”.

There's a lot of different methods, techniques, and challenges that drive incorporation of advancements such as executive decisions, the focus on innovation, and the focus on competitive reactions. However, as FSBO’s don't traditionally address not only the holistic requirements for transformations—supply chains, complex digitization, marketing, staffing, and partners—but the nature of their BAU work contributes to their inability to dynamically make changes as time, results, and competition change the original delivery designs and requirements. When we factor in the regulators, regulatory compliance mandates, and projected industry changes, the need for the framework outlined in the subsequent pages cannot be discounted.

A Deeper Dive: Regulators Throttling FSBO Innovations?

The word innovation as an identifier or action has become nearly meaningless during the last decade—it has become a marketing word more than a reality of intent or realization. When a consumer researches financial firms and their products and services, the word innovation is liberally sprinkled across materials, media messages and websites not to mention content pushed down to mobile users to strive for attention. Merge in the speculative money ($16 to $25 billion USD) being “invested” in FinTech startups and the landscape becomes a peppered mosaic of modern piecemeal solutions. It is all quite confusing and bewildering for anyone evaluating or assessing differences.

The topic gets even further opaque when you factor in the ideas surrounding responsible innovation now being floated by financial regulators. It is an emerging set of principles, ideals, social programs, and disciplines. That is honorable—but also plants the seeds of uncertainty and subjectivity at this stage of FSBO transformation.

For FSBO's already operating with internal innovation centers of excellence, they would likely within these boundaries, add time, expense, and delay market offerings if the supervising agency is unfamiliar with the solutions or the crossover porting of an existing offering. For regulators, this means that they will need to individually or collectively establish their own labs, experts, process, and practices and approval mechanisms to vet financial technologies that are accelerating faster than even the internal corporate groups can assimilate. A tall order for any agency in an era of austerity, let alone enhanced global competition and changes.

Factor in the intense need for privacy and security as part of cybersecurity solutions and the expertise, budgets, and implications of approving or denying any FinTech offering will require an investment measured in the millions if not billions.

For FSBO's to accept that the pace of financial technology will be throttled by an agency that implicitly admits they need to adapt and educate themselves leads to the question across globally interconnected financial supply chains of "what unintended consequences" will be implemented putting domestic firms at a competitive disadvantage and higher cost basis?"

There is much to be discussed in these types of solutions. However, FSBO's, their leaders, politicians, policy makers and the public need to weigh in on what if any responsible innovation they are willing to accept and by what rules. A failure to not vet the vetting ideals or to place shackles on FinTech before it even has a disintermediation impact is hardly capitalism.

Introduction into the Framework

There are seven functional categories in our framework model; 1) data, 2) operations and back-office, 3) channels and markets, 4) customers, 5) technology, 6) competitive pressures and alternative providers, and 7) relevant innovation, or as we mentioned above, responsible innovation. It's worth noting that items one through five are large when it comes to in control of the internal organization and its partners. The last two categories are outside of our sphere of influence, outside of our control directly, and as a result, can have significant influence on delivery and direction.

Additionally, when executive leadership and teams apply their operational components to the seven categories, they frequently have a view that they want to execute tasks and plans based upon historical methods. Yet, new deliveries and rollouts don't follow the traditional paths, especially with items six and seven, as changes quickly transform and "their" critical path is not "the" solution (see call out box below introducing the framework in greater detail).

To plan it all out, to think you're going to know everything before you begin, and to have it all laid out in a budget isn't reality as experimentation become the rule. This is due to the rapid cycling, transitory behaviors of how these products are utilized by fiercely independent consumers. The flipside of not controlling every portion and component of the budget (i.e., not allowing for flexibility as underpinned by delivery via flexible budgeting methods needed during the development and delivery cycle), is that any type of transformation work needs to perform in lockstep with the markets—for on-demand rapid cycling.

The main topics we discussed in the first nine chapters, from commoditization, the supply chains, to looking at the full spectrum of how we deliver, and even how we tackle projects differently, are all components of next steps and the framework for action, execution, and post-implementation modifications.

That's what brings us to Chapter 10, taking all the previous chapters into consideration. What we have put forth to the FSBO is that we can no longer just look at markets, consumers, IT, and so forth from a traditional, serialized, or even a parallel approach of delivering projects which are defined by the consumers (pure play regulatory issues for the most part still can utilize the old). As we have stressed throughout and in the framework below, we should do this through a series of partners, and that could be outsources, that could be FinTech companies, or even with skunk work efforts within our enterprises.

Additionally, it could be launched or lifted and shifted from academic settings. For example, one of the things that we saw recently was that different academic institutions, Wharton, MIT, some of the other big schools are launching a nascent FinTech curriculum for teaching prospective MBA's and those wishing to make a career in finance, how to take advantage of FinTech ideas and principles. It's changing the fundamental idea of how we go about, not only identifying those projects or those initiatives that we want to put on mobile apps or to get in front of the consumer, but how are we going to measure them? What do we have to do? How long do we do it? Do we do it more of an agile approach where our delivery techniques are a time box? Do we do that through partners and whitelabeling?

It's a whole different process of saying, "How do I define financial services to my customer?" versus saying, "I need to control everything the customer does." The next major challenge would be to recognize that financial data, financial solutions are no longer "the" custodians of the consumer's financial profile and probably haven't been for several years, due to the advancement in data including those encompassing machine learning.

The real challenge for forward thinking FSBO's is how do they want to be part of emerging financial solutions, whether it's for clearing, whether it's for difference value-added services, whether it's for competitive linking of information together so the customer has a common profile, common repository, and anticipated needs. These operational principles and methods for delivery are foreign to some of us in FSBO's, because as bankers, as conservative individuals in the finance industry, we define the products that in history we've changed or created.

So, for us to come and approach is differently by saying, "We don't control all the channels, we don't have a lock on financial data, we don't control a complete view of how the customer conducts their life, but we want to participate in those channels, those products, those solutions", is an admission and an acceptance that will be very difficult in dogmatic cultures and mega-FSBO operations.

It's a distinctive mindset (implicit within our presented framework) of how to approach and deliver with market demographics because it also sheds and spreads the varied risks around that many organizations, especially traditional financial institutions, are not comfortable with having these in some else's wheelhouse. In their mind, they should own it, they must vet it, they must prove it, and the only way to do that is it's got to be in-house, or it can be with somebody governed and controlled through a solid set of contracts. Additionally, any solution set must contain robust and highly specialized regulatory capabilities. Whereas the needs don't change, it's the go-to-market supply chain that is fundamentally being rewritten.

Additionally, as governance and compliance mindsets change across FSBO's, what is going to constitute regulatory oversight? Are we going to go back to the principles of the 1990's, as part of a lifting or rewrite of Dodd-Frank, where the "spirit" of the regulation is more important than the "letter" of the rule? That then leads into a whole question for FSBO's of how to get it done and should we compete against a wide-swath of unknowns. "Do I need to be in this business because it comes with a set of restrictions and oversights that I may not want, as terms of governance in delivering 'that' product to the consumer market opens us to exposures or losses?"

It's a fundamental shift in how FSBO's move towards markets, consumers, competition and their own operations—more than a fundamental shift in the actual techniques themselves. Sure, the techniques are going to change as technology changes, so the delivery mechanisms, the vetting, the solutions and the independent verification must be all different. We have dealt with consumer, regulatory and market movements for years, and years, and years, but we've controlled all these. We've had our fingers in and on the pulse of delivery, yet the problem is in the future for many of FSBO personnel is that is not the way the operating guidelines work moving forward. It really coalesces around an adaptable framework that can be used as a go-forward guidance and filter to ensure consumer and innovation relevancy.

A Deeper Solution: A Framework for the Future

Throughout this book, we have talked about the rationale and implications of FSBO markets, operating structures, personnel, outsourcing and the likelihood that with "*Bricks, We're Never Coming Back to This*". And with all the changes afoot, the real demand is to eliminate misguided axioms and focus on the future—transformations, M&A's, partnerships, retirement, product mixes—using competency lanes mapped around centers of excellence (COE).

Whereas the complexity of the approach can be on the surface intimidating, the general foundation for improvement cluster around seven distinct poles—innovation, competition, data, customers, operations, markets and technology. A common trap for most organizations is that their efforts are concentrated singularly around innovation and competition. The former is something they believe will lure new customers or improve profits, and the latter in response to external pressures resulting from other's initiatives. Yet, over the last decade, most corporate efforts are concentrated around these two poles of (re)action ignoring the five other areas that are within the FSBO's control.

For FSBO's, unrealistic frameworks promoted to incorporate the external drivers of innovation and competitive pressures into their going-concern environment resulted in a very high failure rate—that is using a prescription model that neglects the side-effects inherent within populist delivery approaches.

Moreover, because the internal COE work is not as glamorous or prestigious as external events (i.e., everyone wants "innovation" on their resume), the competency levels needed for adoption and adaptation are made apparent only after bumbling and stumbling to failure as items deemed low priority became critical path. Organizations also erroneously internalize requirements that everything must be in flux at the same time rather than segment and iterate current state people, processes and technologies (i.e., compartmentalized in the figure below).

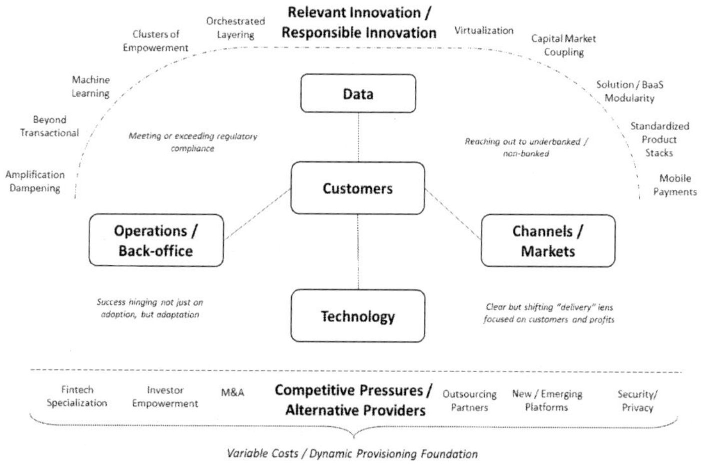

Too often, FSBO's and particularly those within the consumer space, ignore the foundational items needed for success. There are others who take it to the extreme by trying to control and vet all elements of competitive pressure and innovation before they can get anything put into motion—believing in an old waterfall axiom that design is everything even when you are operating in an environment that is outside of direct influence.

Breaking Down the Framework

Looking back over all the issues and advice we've discussed in this book, all the challenges and all the responses to those challenges, you might be asking, "Is there any particular area that is going to be more critical than another for FSBO's?"

We think a critical area, which is going to become mandated core competency, is how do you become the orchestrator or the person who puts all the pieces together for the solutions that paying financial customers are wanting and seeking? That is really something that, be it external, be it internal, something that FSBO's aren't necessarily historically strong at securing a critical mass let alone delivery ecosystem, which satisfies go-forward plans.

You see a lot of integrators, you see a lot of consulting advisors, you see a lot of outsourcers that have that expertise, but within the FSBO community themselves, the people that they rely on internally to help make those decisions, to vet those decisions, being an orchestrator of those decisions, services, products, solutions, product stacks are outside the FSBO world. It is going to be a core competency because of the fast-cycling nature of products and services.

Let's be honest. The customers want things that are tailored to their environment, offerings that mean something to them. You (i.e., FSBO's) don't want to be part of the 75% of the FSBO's that are viewed negatively. You don't want to be part of the 75% that the consumers "don't want anything to do with because they add no value."

If, as a FSBO, you have a culture that says, "I'm just going to do one thing, and I'm going to do it well, and I'm going to build it, and they're going to come," then guess what? That's not about being an orchestrator, it's about being a solo performer in the middle of a symphony playing from a different score of music—you are oblivious to what is transpiring all around you—and it shows.

We've seen a completely different approach when it comes to FinTech companies, and this is some of the justification why the regulators are starting to review the FinTech world, because these upstarts[c] have a nimble mindset. They adhere to the ideals that, "We don't really care if we created it, but we can license it, we can compartmentalize it, we can create a stack of all these components to come up with a value-add that is something new to the finance industry that people will pay for." That's how these specialized providers got their funding and why their efforts and apps are gaining greater shares of provisioning. The approach of assembling stacks, being part of something larger (not launching copycat competitive offerings) has not necessarily been a core expertise within banks, or any kind of financial institution. In summary, FSBO's are responding to competition and the consumers with methods, techniques, and approaches already outdated and not aligned with the seven taxonomies below of how, where, and why.

A Deeper Immersion: Breaking Down the Taxonomies

The reality is that when each of the taxonomies are broken down into more digestible line items, a path map becomes apparent as you work from the growth to the goals to the results (see figure below).

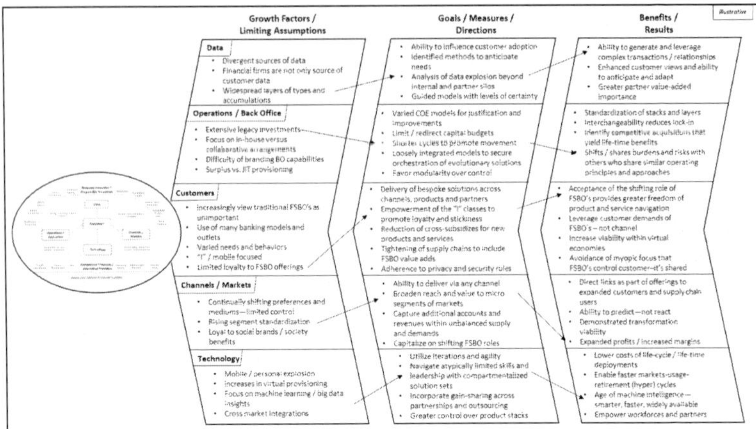

The idea of using lanes of improvement corresponding to required COE's provides iterative focus in what otherwise may be deemed to impractical to accomplish within set time frames (e.g., time boxed delivery).

Additionally, using this framework spurs questions on "What's Next?"

- Where will boundaries be created by customers with regards to the use of their data as accumulation grows and methods of collection blur boundaries of ethics?
- How will banks adapt to a reality that customer engagement will be controlled by channels outside their venue and the customer profiles once deemed mandatory are separate of their offerings?
- How will non-traditional customers be served in a mobile banking-as-a-service economy where loyalty incentives are limited?
- Where can FSBO's capitalize on investments in a global push to match lending needs directly to borrowers and risk-willing lenders (beyond a deposit to lend model)?
- Beyond disintermediation and commoditization, how will traditional institutions compete with alternative providers and platforms?
- Can FSBO's hope to drive change or will they be regulated to behind the scenes plumbing (if they cling to their history)?

- With pressures on spreads, channels, customers and products where should FSBO's be seeking opportunities and how should they build, buy or partner for it?
- With "lanes" of delivery increasingly being specialized and defined, where and how should FSBO's prepare to adapt—or divest?

Using these illustrative questions to focus the framework presented above, we can then objectively examine why the idea of innovation draws the attention and money of banking executives even though they are not able to maximize they benefit of their deployment.

If we utilize a framework to focus innovation deployments, we can cast innovation with acceptable expectations, realistic priorities, and the critical success dependencies. Challenging innovation is warranted if FSBO's hope to achieve their place in future markets and consumer wallets. Innovation must be relevant and responsible—not a cliché that can be latched upon in the hope of defending the castle.

Focusing in on the Architectures—Over the Infrastructures

For FSBO's, a key component to incorporate the above framework is there needs to be a very strong architectural idea of how to assemble things from technology. They must have very good filter and retirement process, but also dynamic capabilities that can expand and contract as needed. Not only for the manual solutions, but for the initiatives that we will spend months and months defining and re-engineering.

FSBO markets and supply chains are increasingly fast, nimble, and quick, requiring architectural leadership over infrastructural management all to achieve orchestrated solutions and compartmentalize reusable building blocks (i.e., avoiding the technology trap of everything in its silo). Existing leadership is not accustomed to competing, let alone ensure upstream and downstream variations across the financial supply chains, with new entrants and unique offerings. Additionally, it's also going to be a material challenge for the regulators because, just as we've seen with cryptocurrencies and other consumer financial advancements taking place across global markets, consumers are going to flock to crypto-solutions (i.e., hardware, software, services, products and processes) whether the governments are ready for them or even whether the FSBO executives are ready for them.

Ask yourself, why would financial consumers say if major e-commerce companies were banks? Market research says they'd switch (and you can better believe these firms would provide them with streamline migration services to move their financial profiles onto their platforms). This is because they (customer-driven cultures) understand the consumer and they have strong architectures that are continually be adapted as new (technologically advanced, layered components) solutions are introduced. They understand how to build these infrastructural stacks regardless of their invention origin (providing it is secure, auditable, and reversible).

We can look across our FSBO silos and examine why it is projected that one-third to two-fifths of all brick and mortar retail establishments are supposed to cease operations in the next five to seven years? They're supposed to close shop or be bought up by online competition focused on not just behaviors, but distribution, supply chains, interfaces, partnering, and offerings. The acquisition of Wholefoods by an "upstart" such as Amazon points to where the traditional markets on heading across nearly every consumer segment. Gone are the business models that believe, "Hey, I've got products. I've got to have storefront to reach my consumer. I've got to do this because this is 'how it is done.'"

A similar set of scenarios are playing out within the various FSBO market segments. However, most FSBO leadership continues to act as if an outside (i.e., non-regulated) suitor would not possibly be allowed (by the same regulators they want to go away) to complete an acquisition. As FSBO's continue their lobbying efforts to say, "We don't want that. We want this..." they are missing what has made others outside the FSBO silos successful.

If FSBO's continue with that same process of how they approach projects, not understanding orchestration, not understanding layering, not understanding compartmentalization or how to be the conductor of a symphony, then guess what happens? Then those FSBO's are probably going to be relegated to what we would call plumbing duties, traditional banking as a service, functions contained in the back office that don't add value. In this scenario, the FSBO offerings become a commodity and their business delivery becomes all about price sensitivity.

The people that are the true conductors using architectures as one of their main tools for orchestrated transformations they're going to be the teams and entities that are leading the transition because they're going to be creating solutions that consumers want to use and pay a premium for it. These leaders don't care whether that was a FSBO channel once upon a time, or that was an electricity meter-reading channel for an app that ties directly into their bank accounts. They see opportunity.

A Deeper Digitization: Why Use the Framework for Complex Tasks

Reinforcing an earlier point, what arises from a comprehensive digitization analysis is that there should be a conductor or orchestrator for all enterprise digitization efforts. Moreover, depending upon the size of the enterprise and geographic spread, the responsibilities may fall within a cross-functional team (i.e., leader team) that is accountable for timeliness, decisions, exceptions, integrations, programs, and of course, architectural fit (e.g., product orchestration).

Stated differently, the incorporation of new and emerging digital solutions into an on-going FSBO requires that they are part of a cohesive mosaic, a journey map if you will, which vastly improves the economics and adaption of internally or externally vetted solution sets.

Treating the perpetual demands of digitization as a one-off project to implement and forget (the paradigm of a sponsor approach), fails to account for changing consumer behaviors, rapid cycle technologies, and advancements in software and P2P media interactions. In a FSBO, a few axioms for digitization program initiatives include:

- Stitching hardware, software and services layers requires forethought, afterthought, skills, and commitment.
- Connecting and automating digitized workflows requires a cradle-to-cradle culture.
- What is not always a formula for success is that, "if you build it they will come."
- "Me-too" strategy and architecture adoption will yield, at best, commodity-like returns.
- Domino effects need to be anticipated and incorporated into larger programs.
- Data digitization is like an amorphic sand pile—one grain in a piece of a larger solution that until mixed with agents (i.e., applications), is still just sand.

So, we have a collective industry that is not only faced with a contraction and being pushed by digital advancements from competitors, but by the very consumers being sought out as "prime". To survive and hopefully thrive in a "world gone mad", the use of orchestration (and its close cousin's encapsulation and compartmentalization) beyond decision making must now be extended into strategy, design, development, implementation, infrastructure, revitalization, and even the very corporate cultures that seek to gain advantage from their deployments. Why? The reasons are long, but a few of the salient rationale for our purposes include:

- Criticality for cloud services and deployments.
- Enlighten, enable and stimulate adoption and adaptation of rapidly advancing digital solutions.
- Composite assembly into layers where improvements can be made without impacting the entire stack (s).
- Scale up or down as demand and budgets allow.
- Building blocks for greater solution sets and able to provide for reusability.
- Deliver "old core, new wrapper" solutions to leverage or bridge legacy operations and applications.
- Feedback and monitoring to "Think BIG, but start small" foregoing large capital outlays, while honing market features and functions.

As the FSBO digitization intellectual capital shifts and multiplies, so does the confusion of which representations should be selected. Whereas, selection of strategic development frameworks often boils down to corporate sponsor preference, inquiries should be made as to what is the benefit for the culture, operating environment, and delivery intent before chasing after a prescribed model.

- Which digitization approach should be used without breaking the budgets?
- Do we have the skills to do most of the work ourselves or are we beholden to external expertise (e.g., consultants, outsourcers)?
- What discrete measures are we trying to book and

by when?

- Do we invoke stepwise sequencing?
- How do we ensure success within markets and channels?
- What tactics will bring the rebuke of regulators?
- Moreover, if we fail, does our corporate culture allow survivability for those involved?

There are dozens of additional questions and assumptions that should be addressed before any viable digitization solution set is attempted (notice one missing, "Has it been done before?").

Moving Beyond What We Know for Sure

We know it that the ideas presented within this book can seem incredibly overwhelming, but not if you're an orchestrator. If you had to go in and do it using the old approaches and cultures, it would be, but as an orchestrator, you can do it.

You typically don't write an opera when you've never written a refrain. You don't write a symphony is you fail to comprehend a piece of music. We're not all Beethoven's and a sure recipe for failure is to think, "Oh, I can do this overnight, and I'm going to create it." Guess what? You're not. You're going to do this in steps. You're going to execute against the framework in pieces. We don't think any of us are that gifted (let alone our institutions) when we are looking for new solutions, new profit alternatives, new partners we can magically look at the breadth of actions needed, at the same time creating a roadmap to the future at a point in time.

We don't look at things as orchestrators within and across FSBO's. That's not who we traditionally have been, but that mindset, the cultures that are within our enterprises, that's what's left and that's what always gets us into trouble — "We know what just ain't so." Therefore,

FSBO's do such a great disservice to their customers, and why they run afoul of regulators—all the while being surprised they are under the heel of enforcement actions.

We think you're also seeing that in published customer statistics and infographics from groups like the ABA (American Bankers Association) to the regional and national MBA (Mortgage Bankers Association) indicating, "Your customers see you this way." The FSBO's when they are informed of divergence from their own self-perception respond with, "We don't see ourselves that way. We thought we were doing a better job. Are you sure this applies to us, we're different?"

This difference between market and self-perception is because WE KNOW we're structured for delivery success, WE KNOW our measurements and dashboards tell us how we're doing, WE KNOW we meet everything in the framework within this chapter and can back it up with lanes of competencies.

For some FSBO's this is an important shift about how we deliver and primarily perceive our own capabilities. If we take advantage of what we do good, take advantage of the things that we are skilled at, adept at, and have core competencies in, and map those to "We're going to be an orchestrator of all these things," we can then begin to advance quantumly versus incrementally. We think that is going to be the real challenge, and that's, in fact, what the framework provides for delivery in terms of what's next. It aids with the exposure and correction of misguided beliefs and dispels those items where WE KNOW, to a framework where we understand and improve. This is how an orchestrator operates.

You approach things differently as an orchestrator. You approach things differently than if you were an innovator. Innovators are typically looking for one big shot, like a cannon going off hoping to win the battle in one stroke. They want to rocket to the moon. For example, what we need as orchestrator are small flights going from regional airport to regional airport. We don't need to touch the moon to go from Cleveland to Milwaukee. That, oftentimes is what happens is that we overshoot the mark because that's what WE KNOW we must accomplish, and this is how we traditionally have designed and delivered systems lasting decades.

We design big, foolproof, failsafe solutions that by the time we bring to market, we're playing catch up, are over budget, and end up alienating (or underwhelming) the very demographics we sought to capture. We look at all the security, privacy and risk along with everything else that are mandates, and we think we should be the ones that create and control every piece of these specialties, every piece of that application, and be the ones who have crafted every line of code, every use case and every feature function. It becomes narcissistic, and as you can surmise, a "white elephant" approach that ends many careers (even though it makes good resume builders).

Yes, you must do your due diligence, but as an orchestrator, we don't sit down and put our hands on every department, every individual, and every performance measure—that's why set success levels and requirements so that we can delegate and objectively conduct IV&V's (independent validations and verifications). We rely on individual, team, and partner expertise, their training, their skills, their capabilities, and we can offer them improvements based upon what we're trying to do as part of a program of work (i.e., any projects). Analogous to performing against a musical score, we are the conductor. We are not people within the audience or on the stage who are contributing to the ambience or mechanics of performing. As the orchestrator we follow a framework, we assemble a group of people who are the best of the best. We're striving to make all participants perform with minimal oversight rather than be micromanaged—when we are trying to be agile and aggressive against a plan of action.

Our solution, customer centric framework is about asking a different set of questions, and approaching our challenges from a few degrees beyond BAU. For instance, when asked about a customer fulfillment solution, an employee in the FSBO might say, "Well, we're following best practices," and we would ask, "Why do you always need to create best practices within your enterprise? Why not partner with somebody who has the premiere solution set and create even something better to evolve 'the offer' discussion?" The mindset of executing against what we know, is often why we fail to meet expectations internally and externally. It's a change of how FSBO's approach transitory, stepwise improvements and transformational initiatives—this is where our framework excels, and isn't what it is about?

A Deeper Reality: It's What We Know for Sure That Just Ain't So

It was back in 2009, at the height of the government releasing $700 billion in TARP money, we were driving in a suburb outside of Cleveland, Ohio when noticed a recently closed Donato's Pizza with a signage overlay stating, "Coming soon, your new banking branch to serve you better". Really, another banking branch on a street where two others were just closed and four others still exist?

Even eight years ago, physical branches were transforming drastically because of smart phone emergence, consumer behavioral shifts, and an economy that had leveraged itself beyond its ability to repay. As we shook our head in disbelief on the reasons not to expand a brick and mortar (B&M) footprint in a deep recession, we were reminded of Mark Twain's quote about reality—"What gets us into trouble is not what we don't know. It's what we know for sure that just ain't so." (BTW, the branch was closed a few years after it opened.)

Now a decade later, it's a cycle we seem to be on the verge of repeating with expanding AI and VR solutions, the demise of B&M retail, and with household debt nearing $13 trillion (spurred by the trillion dollars rise in student loans). Today, this is where Financial Services and Banking Organizations (FSBO's) find themselves—struggling with the past, while trying to comprehend where to direct precious funds for future growth and profits.

Yes, there are many leaders who still cling to axioms that have long since expired—unable or unwilling to see the vast changes taking place from consumers to technology to widespread commoditization all within one of the most regulated industries in the world. Many of these leaders are surrounded by staff of their own choosing, consultants who rely on project payments, and even investors who demand staying the course over transformational boldness. It's a quandary to be sure—but the alternative is to recognize too late that "what we know for sure that just ain't so."

As examples, before the Great Recession of 2008, we knew the breadth and depth of FSBO staff and partners surrounding all aspects of financial products and services would create a new utopia of leveraged profits. Before the recession, we knew investments would appreciate with little downside risks. Before the recession, we knew the safeness and soundness of the financial systems—global and domestic—were unshakable and likely never come into question. Before the recession, we knew many things for sure—but that is what got our stoic FSBO's into trouble, while following the Davos celebrities off a cliff of their own making.

In revisiting this popular Mark Twain (i.e., Samuel Langhorne Clemens) quote, we started in 2013 viewing FSBO's and their operations from a world that was no longer of their design and farther from their control (i.e., we saw the reality of Dodd-Frank fracturing their grasp). For you see, their design and operating beliefs have changed only incrementally, compounded from the mid-1960's, even with the abundant adjustments from technologies, processes and regulations.

What was substantially and consistently missed were the consumer changes and attitudes toward FSBO's especially as empowerment in purchasing, social interactions and brand voice rose beyond the brick and mortar of the once valued branch (only deemed valuable by a dwindling Baby-Boomer generation). Bankers believed that with efficiency and technology they could stay ahead of a curve that they strategically envisioned—because they created the world they wanted to change.

It's About Being Humble

Orchestration has been characterized as "holding a wolf by the ears—you don't necessarily like it, but you dare not let it go.[ci]" To find the right solution for each enterprise requires less about interpreting a crystal ball or exerting conformance willpower, and more about business and technology orchestration against a stretch-goal roadmap under the guidance of a "leader team" using a common framework (which should never be confused with prescriptive methods).

We recently attended a sponsored event where a futurist spoke about the "inevitable" changes in FSBO business and society brought forth by advancements in technology and data availability. At the end of this quite thoughtful speech, we came to the realization that the future is a necessary fool's game—especially in a regulated business that must succeed at the day-to-day. The future isn't prescriptive, its adaptive, flexible, and with a framework approach, the ability to make mid-course corrections is easier and less costly.

Within our FSBO operations, iterative change happens rapidly because of innovation sprints. Plus, developing a roadmap is not the what a futurist typically does, and predicting future innovation needs is not necessarily strategic planning. Meaning, that to steadfastly implement FSBO advancements as part of a transformation strategy without mapping iterative steps and delivery requirements is analogous to not recognizing construction signs on highways—do it at your own peril.

As an example, believing that everything linked via mobility stacks (e.g., to reach Millennials, the underbanked) inherently provides initial profit without accepting the reality of organizational cultures, infrastructure readiness, transformation requirements, stack integrations, necessary staffing, data complexities, customer perceptions (via marketing), and even regulatory compliance, is fraught with unintended liabilities as an implication of adoption. It's a huge amount of bravado to ignore everything that should be considered within a framework delivery.

Another reason we need to be humble? FSBO's rank in the bottom .001 percent of firm's which consumers hold in high regard. Furthermore, perhaps because of their image, behavior or media coverage, FSBO's are losing talent and failing to recruit quality personnel.

As we know, FSBO's are a collective set of diverse operating segments (e.g., investment banking, mortgage lending, retail banking) that since the start of 2007 appears to be at a precipitous, rapid domestic weakening even thought their profits now approach pre-Recessionary levels. This is also reinforced empirically by lack of organic growth (e.g., M&A demands), declining or stagnant share valuations (e.g., regulations and investor sentiment), capital margins (e.g., regulators), stress tests (e.g., sovereign central bankers), and even after eight years a shrinking employment base.

There are many other examples we could use, but these all point to a required shift in operating philosophies. The regulations themselves have allowed FSBO's to thrive even as they rail against their guidance. The use of our framework is in preparation of what is happening now and what will be forthcoming as commoditization sets in and as the traditional supply chains are disintermediated by those who don't behave like traditional finance personnel.

A Deeper Reality: How Will the Mighty Fall?

For banking leaders, change across channels, markets and customers will be fast and uncompromising. To take time focused on strategy or approach to the detriment of delivery and results can, if not framed successfully, a set of career-ending delays.

Yet, to find a better approach or framework to ensure that all elements necessary for success are identified and delivered, moves consumer adoption and satisfaction beyond the statistics where 75% of banking customers firmly believe these institutions offer no value-add to the evolving financial supply chains. It is still the ghosts of our predecessors walking down their mahogany corridors reinforcing what "just ain't so".

Will it end with FSBO's creating vastly different financial supply chains and applications? Will it end with banking-as-a-service models pushing traditional providers into the infrastructure and whitelabeling industries? Will it end with vast M&A actions—with the mighty pushed to the future fringes?

Strategic, Tactical, and Daily—Don't Lose Sight of Each

As you've probably noticed, as we wrap up the final chapter of this book, is that we've been moving between strategic, tactical and daily demands without ever getting hung up on naming them. It's a vastly different approach that those practicing strategic views (i.e., the management consultants) and those practicing tactical management (e.g., integrators and project teams) to those dealing with the day-to-day (i.e., employees and outsourcers). Our framework focuses the previous nine chapters by bridging gaps between the strategic, tactical and daily. It is our opinion, when reviewing the FSBO industries, that they are overwhelmed because they are not getting it done (i.e., closing the gap on non-traditional competition), nor offering what consumers want.

What we offer FSBO's is the fact that it doesn't take an army to fix problems and adapt to challenges. It can be taught. It can be learned. It can be educated and improved upon, and that's what an orchestration is all about. It's not about, dare we put a sacred goat under a knife, the high-priced prima donnas who are brought in to turn the operations around, embark on new directions, or to make the sky's rain profitable demographic segments.

The FSBO's always have these within the various business units. They have somebody doing loans. They have somebody leading the trading floor. They are the leaders and no one dare challenge their operating directions. Everything revolves around them. As conductors (who are part of an orchestra), we kind of look at that and say, "It's not all about me. It's about the people I surround myself with." FSBO's will never survive with the egos that built those foundations—at this point, FSBO's are fast becoming the "new Rust Belt".

And there is another reason why our framework will be foreign to those within traditional, conservative FSBO's. They hire people that are like-minded, and they put people in their staff that they agreed or approved, and kept them there through rewards and position.

These FSBO leaders sometimes surround themselves with consultants who come in and they say, "Oh, yes. You're exactly thinking the same way because that's how we think,"—and you're paying them for this advice. Then you go to the regulators, and they're usually the last of the market that's going offer any guidance on leading edge approaches—they say, "Oh, yeah, that's fine." Everybody gets this groupthink from someone, "Yeah, everything's great," but it really doesn't advance the agenda because we surround ourselves with people that tell us exactly what we want to hear. That's deadly.

However, that's what's happening, and that's how FinTech has really come to be more than buzz and glitz. The importance is that FinTech itself is evolving and highly varied requiring FSBO's to jump seamlessly back and forth between strategic, tactical and day-to-day.

If you look at some of the things that we've been talking about, for example FinTech courses, they can't create a book because it means so many different things. It could cover blockchain technology. It could mean different types of solutions that are on your mobile phone. It could mean a whole host of advancements that include risk management, AI, machine learning, virtual reality—and everybody calls this FinTech.

In the FSBO world, we call everything FinTech that is outside of existing operations, so what do we try to do? That's where the orchestrator comes in and his group of people that are, what we call, sub-conductors (i.e., first-chairs), they get in there and say, "Yes, we don't necessarily need to define everything we do as an FSBO trying to avoid the traps, while being progressive in an effort to serve the customer.

We don't necessarily need to label everything we do, but everybody's got to be on the same page, marching collectively forward to achieve a known set of goals and objectives. What we need to do as the conductor is to get the best out of everybody that we solicit value from."

That's what a conductor does. It makes people better, by extracting the best each must offer within the roles they perform, not necessarily makes each member or contributor the same—absolute conformance.

We play to similar tempos, we follow a score and we practice making our results against a common framework more efficient and widely accepted in the markets. We don't want everyone the same, if we do we will find very quickly that the world needs FSBO functions and capabilities. That doesn't mean it's a given that the world needs the FSBO's we have today.

The message throughout this book has concentrated around, "***What gets us into trouble is not what we don't know. It's what we know for sure that just ain't so.***"

For FSBO's it remains to be seen if they can leave the past behind them even in the face of record profits. For money will mask the competitors who will completely outmaneuver cultures steeped in tradition—and not the voice of the customer who firmly believes FSBO's are like ferocious dinosaurs—destined to be fossilized.

And finally, like some who embarks on a journey or fight to implement change and transformation, be careful not to outdistance your teams and enterprise. After all, it is a very precarious position to be leading the charge—only to find out no one is behind you.

Illustrative Challenges / Opportunities

		Controls and Implications (The 4-V's)			
Category	Illustrative Challenges / Opportunities	Volume/ Rate	Variety /Adapt	Velocity/ Freq.	Veracity/ Accuracy
1	Utilization of a multi-faceted framework to diagnose complexity, incorporate drivers, and establish COE that will provide a path forward for FSBO's grappling with where to begin and how to proceed.	↑	↑	↑	↑
1	Deploying a next-step framework that holistically provides for upstream and downstream changes, rapidly aids with the transformation of rigid FSBO cultures not accustomed to cross-industry solution incorporation.	↑	↑	↑	↑
1	Solutions of tomorrow demand an efficient framework stitching together people, innovations, infrastructure, services, partners, and leadership.	↑	↑	↑	↑

Category	Illustrative Challenges / Opportunities	Controls and Implications (The 4-V's) Volume/ Rate	Variety /Adapt	Velocity/ Freq.	Veracity/ Accuracy
1	Incorporation of a framework into operational or transformation improvements, allowing FSBO's to define a mosaic of options and solutions rather than accepting vendor or outsourcing lock-in.	O	↑	↑	↑
2	With new frameworks, new measurements, methods of reward and active monitoring must be developed beyond common FSBO data mining and intelligence.	↑	↑	O	↑
2	FinTech and patents are not the only viable taxonomies for FSBO's to add solutions, change behaviors, influence markets or resonate with consumers.	O	↑	↑	↑
2	Acceptance that traditional front, middle, and back office reengineering is secondary to customer behaviors, security and privacy and responsible innovation within an across channels and technologies.	↑	↑	↑	O

Cate-gory	Illustrative Challenges / Opportunities	Controls and Implications (The 4-V's) Volume/ Rate	Variety /Adapt	Velocity/ Freq.	Veracity/ Accuracy
3	Incorporation of a framework, not a prescriptive methodology, allows for adaptation and iteration to promote compartmentalization and standardization.	↑	↑	O	O
3	Selectively define and deploy mobility to reach new, untrusting demographics is not a panacea to be blindly followed.	O	↑	↑	O
3	Acceptance than next steps along a common framework will be amorphic and transitory is a given—short-sequence, clear boundaries, and rapidly delivered is the "new normal".	↑	O	O	O

↑ – Increasing non-linearly, unpredictable behaviors
O– Increasing but predictable

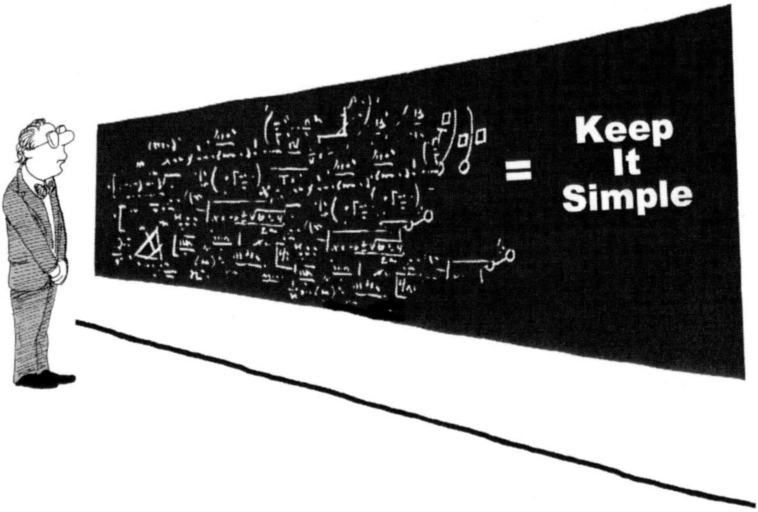

Image ©Shutterstock, Inc.

Appendices

It never fails, when the book is completed, there are additional items that warrant coverage, but if included in the mainstream text, they would lead to tangential rabbit holes. Therefore, we have assembled a brief set of appendices that provide a few areas of foundational depth.

The first two appendices are a subset of extensive research published and unpublished since 2015. Much of the original work has been continually updated and is reflected in these reference readings.

While these appendences can be used as standalone reads, they are segments that enhance the frameworks for desired outcomes rather than getting lost in a plethora of detailed tables, diagrams and interpreted results. We have the details, but they are deep tangent for this topic of **Beyond the Technology Traps**

Appendix A: Digitization Complexities
Appendix B: A Discussion of Big Data
Appendix C: Acronym and Phrase List

Appendix A—Digitization Complexities

(An abstraction from "**Beyond Corporate Project Mentality:** *A Multi-Year Investigation into Financial Services' Digitization of Everything (DOE) Uncovers Why Many Efforts Fail to Achieve Desired Results*", research from Mark P. Dangelo)

Digital Yin and Yang—the darkness and the light. Some describe this as striking a balance of duality between complimentary forces. Similarly, in a world now encompassed by enormous data output and precision decision making, there are emerging taxonomies or disciplines which are balanced by privacy versus profitability, marketing versus identification, protection versus accessibility, accountability versus irresponsibility, and fraud versus ethics (i.e., digital integrity).

When these discrete continuums (along with others) are assembled, they conjure a "picture" of how particular digitally driven enterprises operate utilizing a cultural profile to achieve results and repeatability. Simple enough, no? No. With the speed of digitization, nothing is as it appears and its complexity (along with hidden implications) are anything but straightforward especially for organizations unaccustomed to rapid change. The light-speed of digitization is anything but BAU.

Yet, it is the assembly of this puzzle of these complimentary and competing attributes that make digital adoption (and subsequent adaptation) so difficult especially within highly regulated industries such as financial services. Underpinning this evolving digital picture are the realities of organizational capability, acceptance and readiness—aka its culture and complexity of interworking.

Since early 2013, I embarked on a professional journey trying to ascertain what is the secret sauce for digitization success and longevity? For nearly three years, I've assembled hundreds of artifacts and thousands of pages of published research, case studies, and recommendations spanning academics, consultants, vendors and thought leaders.

However, as I attempted to assemble a common go-forward rationale, I realized that few ideas fit together and little of the solutions spanned more than a narrow range of applicability, or were just a macro blanket encompassing digitization inevitability. Additionally, I believe and contrary to current tenets, that digitization may plant the seeds for the next great "depressional" shift as cross-domain workforces, consumers, investors, corporations, data and of course, technology permanently transform trade, purchasing, supply chains, and an ability to repay (or participate) in any economic interchanges.

Digitization, the very thing that makes processes faster, data increasingly accurate and pointed, and consumers clamoring for more, could provide the accessibility for foreign wars, cyber terrorism and unintended consequences that cannot be solved with central banker's "extraordinary powers". With a decade nearly gone since the start of the Great Recession and ensuing financial meltdowns, could we fathom how fast economic collapse would come with the hybrid cloud solutions, data gathering, system interconnectedness, and consumer power available with just mobile devices?

To adapt and utilize unchecked digitization solution sets (where digitized data is moving from 22% usable to 37%), pundits are increasingly recommending cross-industry operating model for existing financial services firms—incorporate innovation incubators, create software firm cultures, utilize an Amazon-like approach to serving customers (i.e., retail). It might be concluded from the recommendations that to be a viable FSBO firm, you need to operate as anything but a FSBO enterprise.

Yet, if we want to survive the industry onslaught, digitization beyond one-offs must become the norm. If we want to move beyond regulations (as we know today), we should leverage technology via interoperable stacks. If we want to provide journey maps for corporate onboardings, communications, orchestrated methods and techniques (based on Agile), we must replace the hodge-podge of solutions from vendors, outsourcers and our own staffs.

Within our FSBO operations, iterative change happens rapidly as a result of innovation sprints. Moreover, developing transformational strategy is not innovation, and predicting future innovation needs is not necessarily strategic planning. Meaning, that to steadfastly implement digitization advancements defined as part of a multi-year transformation strategy is analogous to not recognizing construction signs on highways—do it at your own peril.

What has resulted from our criteria analysis is that digitization efforts (in-house or outsourced) can only be effective if placed into a framework of interconnected requirements and evaluations—and not taken out-of-context or in isolationism as part of traditional one-off projects.

Think of it this way, much like the debate over driverless cars are being thrust into the front pages (where traditional automakers caught unprepared are casting the most doubt on the upstarts of driverless cars), FSBO can learn from the mistakes and perceptions playing out in another industry. Driverless cars indeed are about using "parts" from many disciplines to arrive at a new solution (i.e., orchestration).

For FSBO's, our historical ideals are getting grey—it is time for "new leaders of the band". We could be the auto industry fighting a consumer change—is that the position we wish to take? I for one am not willing to just let FSBO solutions go to new entrants. We are in an age of FSBO innovative enlightenment—long live the Digital Renaissance.

"So, what" you may quip, much of the world has begun its digital transformation[cii]—several are undergoing second and third iterations. Illustratively, in 2014 the number of cell phones surpassed the number of people on the planet, and in a few countries, there were nearly two cells phones per capita. Smartphones (e.g., Galaxy, iPhone, Fire) by end of 2015 will total 2 billion units—not including tablets or reading devices (e.g., Kindle).

Additionally, digital payments are gaining significant shares of consumer purchases brought on by "generic" payment methods (e.g., Apple, Google, Amazon) as well as consumer apps (e.g., Starbucks). Even autos, the second biggest "investment" made by consumers, are increasingly tied into a digital world in areas of performance, information gathering and sharing, roadside assistance, personal electronics, entertainments and of course, shopping.

It seems the idea of the Internet of Things (IoT) knows no boundaries, and by using digitization, automation is being transformed into actionable "intelligence". Consumer behaviors are being altered at varied rates across demographic segments, thus changing rules-of-engagement channels and downstream forecasting models on predictability.

Taken in its entirety, what is coming into focus for financial firms is that those who have the underlying data and digitalization strategies will prosper in the transition from automation to digitization (i.e., the second and third waves of business transformation respectively). Challenges, in their simplest form, focus on prediction, intelligence, protection, growth, indicators, profitability and most importantly digitization relevance.

They must answer to investors, consumers, public, regulators, politicians, activists, peers, and the list goes on. Financial firms are sensitive to risks and untested approaches even though consumers are expanding their use of digital apps, business solutions and of course, service-delivered solutions (e.g., Uber).

Digitization is, if looked at holistically, a primary reason for compliance challenges (e.g., TRID), a key factor in determining future credit risks, forecasting consumer behaviors and a bullseye on the challenges with restarting private securitizations (and transforming the GSE's). Yet, in our quest to chase the Holy Grails of "All Things Digital", the "best and most profitable customers", and "operations which exceed the expectations of regulators and investors", have we embraced digitization solution sets simply because it is digital (e.g., big data, social media)?

For the remainder of the decade, anyone in a FSBO should expect M&A's driven by evolutionary digitization. In fact, it is estimated that up to $300 billion per year will be minimally raised for M&A events many of which are driven not singularly by products or services, but by connecting with customers, greater social outreach and access to information for prediction.

The temptation to expand digitization to “know more”, “do more”, “be more” will increase as consumer digitization grows. But with continued industry consolidation on the horizons, the rationale for action needs to be measured against the key indicators that are tied to mission, strategy, goals and objectives.

What firms and their leaders need to realize, like those in the stock markets, is that to chase the market during a growth frenzy is fraught with hidden risks and downside demise.

Just because it is digital, doesn’t make it accurate. Just because it is extracted and manipulated across machine learning algorithms doesn’t mean that the result will produce success—lasting or iterative. Digitization is not a substitute for thinking.

What is concerning, especially when people across FSBO mention bubbles, over-extension, risks, and a host of other technological challenges, is that chasing digitization without purpose and clarity is analogous to the market crashes of 2000 (technology) and 2007 (housing). What will be happening before or after 2020?

“We’re dancing” as a prior CEO said just before he lost his job, but are we confident in why and how? Are we chasing it for competitive reasons? Are we chasing digitization because of media reports? Are we chasing it because our employees and focus groups believe it is the future?

All the reasons are valid—if we know why, if we know how, as long and we know by what means the pieces efficiently fit.

Digitization is nothing new (heck it is representative of computer science cornerstones dating back nearly 40 years ago)--its acceleration and pervasiveness is where adaption, integration, organizational, cultural and implication challenges make winners and losers. Digitization is fundamentally about what you do with it—explicitly RESULTS!

Furthermore, how information is gathered, processed, merged, extrapolated, retired, transacted, protected, managed, strategized or anything in between cradle-to-grave regarding the sourcing chains of all financial industry's products and services, is likely inadequate when assessed against transformative performance standards. For the last four years, chasing the goals of digitization (and realizing measured and consistent returns) has been comparable to chasing an apparition.

In a corporate era defined by aftermaths and results against consumer's behaviors and sovereign economies, where is digitization heading? How will it impact our current operations? What are the pathways to embrace for assurance of success when contrasted against where we are today?

For technologists and developers of software, the last decade has been thrilling as the science of data and its generation has spiked new corporate disciplines including machine earning, big or vast data, advanced AI, data discovery, business analytics, market repositories (assembled for identification and resale), machine-to-machine interactions, Internet of Things, sensor science, mobility advancements and efficiencies and the list goes on.

While some professionals label these advancements as (positive or negative) disruptions, risks, outcomes, structural shifts, consumerisms, innovation, or as convergence in delivery, whatever it is called the quantitative high-water marks continue to exceed projections.

In our economies and financial supply chains undergoing a very broad digital awakening, it is interesting to see front page headlines touting, "Increase capital and shelve Dodd-Frank". This news is generated from yesteryear leaders who forgot that if we put the Great Recession back 20 years, the lack of computing and digitization may have let the likes of Countrywide and Bear

Stearns survive.

Yet, with the information discovery and global FSBO system integrations embryonically available at the start of the last decade, losses, risk and unwinding could not be hid (perhaps if they could, buying once leading brands time to recover) as information was readily available. Wish upon wish, we can only go forward and assuming or making a case for restoration of controls forgets the methods, techniques and depth of information available to today's modern investor, lender and borrower.

Why not recognize the impacts of digital availability and its multifold increases and architect a system (and governance and compliance) that meets a future—rather than assume like the Luddites that the IT world has not already begun to shatter Moore's law? And, as we all have been educated, past performance is NO guarantee of future success, so why try and think we can restore a time when technology was less pervasive?

Somewhat out-of-phase from the initial breakthroughs, (re) birthed out of necessity and innovative emergence, come cybersecurity (i.e., white hats and black hats), multi-level (learned) compliance, consumer privacy, data obliteration, digital sources and uses, value measurements (and loss of benchmarking), brand impacts, intellectual rights management, governance and government rules of engagement, rich-media management, storage management, integration and quality assessment software, cyber legality, digital transformation and a hundred others in advanced stages that weren't even out of the R&D labs (or basements) just two decades ago.

Digitization requires a difference in thinking, different skills, and a different attitude. It also requires, in some cases, a destruction (or disruption) of the past. Not everything digital is innovatively relevant to FSBO's. However, as we have seen in the first two parts of this series on digitization, the events are clearly set to disrupt nearly all aspects of financial procedures, products and services. But unlike the past where destinations were defined at the end of a monolithic solution, tomorrow's journey has very little to do with either.

Digitization is a journey filled with vast growth, unstructured solutions and a fluidity that will chafe many established financial institutions. For institutions like TD who are investing in MORE brick and mortar (B&M) locations, you can safely bet that these future branches will be digitally provisioned with nearly every high-tech solution designed to embrace customers expecting more than a simple human interface. They won't be the B&M's of the past or have anything to do with branch only ideals of the last four decades.

Digitization requires skeptical thinking, but in the end, it will create a chasm between those who leverage it embracing its uncertainty and frequent transitions.

And what is more psychology than technology, just because it is Tweeted, posted as a status, pinned on a site, or anything else, does not make it real or even valuable. Digitization brings a whole new meaning to "be careful what you accept as fact."

Some authors have termed an enterprises ability to adapt to digitization as "cyber cultures". With increased frequency, the days where respect driven by office size, organizational structure, and even profitability are shrinking. If we look within our own corporations, we notice that those with access to and protection of information are commanding greater respect without all the trappings of traditional employees who "climbed the ladder". An outcome of DOE (Digital of Everything) is that it has made a shamble of traditional organizational design and brought to the surface that those who command information or its creation are far more in demand than even the managers who still seek command and control. It's this chaos of design and politics that threatens inaction even when the best efficiencies are made available.

Recently Reuters published citing Brand Finance as its source, that nearly $71 trillion dollars (i.e., half) of corporate valuation is comprised of intangible assets across 58,000 firms. What this rapidly growing number eludes to is that corporate cultures will have to be more about information and digitization rather than traditional benchmarks, asset management or even historical communications and investor reporting. While some economists will cringe, consumption and economic prowess is increasingly about information and how it is captured, managed, manipulated, archived, and destroyed. To combat this unintended consequence of cultural chaos introduced by digitization, proactive and leadership organizations have begun to embrace change using emerging management practices.

Within FSBO's, these unintended consequences will be hard to internalize. Why? As banking balance sheets shrink, there is limited desire to veer from the norm and challenge the procedures and hierarchies, which some credit for a return to stability (and reinforced by regulation). In many cases the unintended consequences of digitization are too much to accept—A Queen of Hearts scream of "Off with their Heads" reverberates down the Mahogany halls.

This disbelief of events by some in FSBO is precisely why in a post-recovery world many FinTech startups have found homes with non-banking institutions seeking to break the casts surrounding lending, securitization, insurance, credit scoring, closing, servicing, and of course, capital markets. Digitization for FSBO, like the presidential politics unfolding, is less and less about dynasties or birthrights. Digitization is the future of FSBO, and if the culture is ready, it is a critical factor for an enterprise's deployment, success and profit.

To prove that these ideals could possess efficacy, and of course, how to let organizations profit and adopt positively to the inevitability the models needed discrete components and further breakdowns. What resulted during a yearlong effort for categorization quantitative assignments, are over 7,500 intersection points spread across 14 taxonomies (each with 14-26 sub-categories) and six grouping of digital impact segments (decomposed into 24 sub-impact zones).

Stating the obvious, each of the sub-taxonomies within that data (and the thirteen additional segments) can be broken down further as needed to properly align with the enterprises goals and timeframes.

Moreover, this multi-year effort has shown is that digitization is a moving target that is often and simplistically within time-starved enterprises just associated with technology advancements—or worse bolt-on solutions to merely expand core systems.

Digitization is not, as marketers sometimes ascertain, a lock-in by demographic segments (e.g., Millennials). So, as we move rapidly forward in FSBO with digitization initiatives that are no-longer just turnkey projects, they require especially over the remainder of the decade a realization that banks and bankers are "swimming in the shallow end of a very enormous pool".

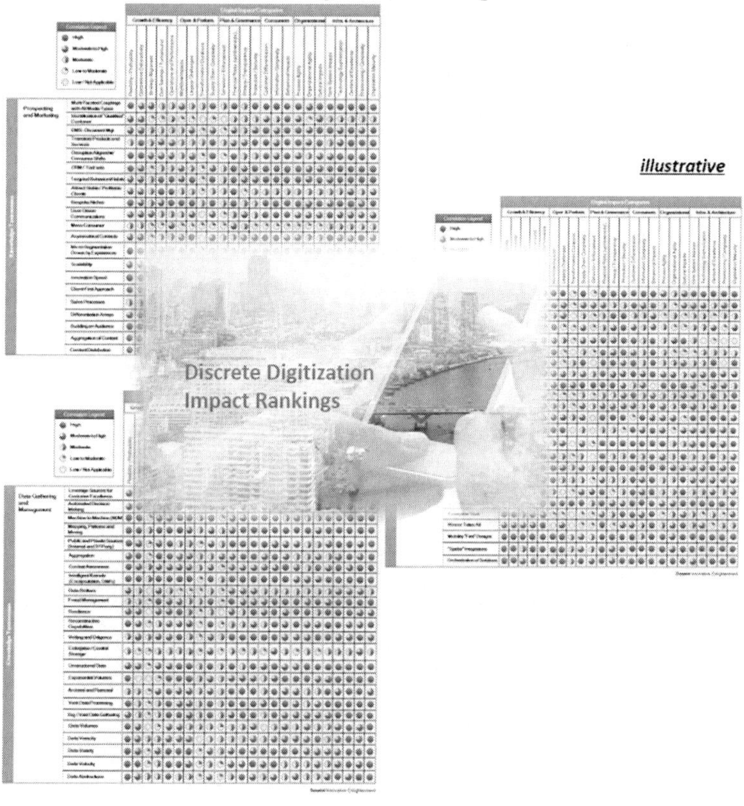

As I recorded these empirically into cross-linked

databases, they generated a repeatable set of processes (and outcomes) that can be custom tailored to an organization across any industry arriving at a "digitization index". This repeatable model is not only advantageous for current and future states, but for surveys and custom generated IC that could be conducted with a "mobile first" approach and integrated with big data sources. Hence, it is a model that adapts to time and changing circumstances, while moving with firms rather than simply at the bequest of their high-priced advisors.

With the development of the models and criteria it reinforces that digitization is a journey—not a discrete destination or project to be completed and then checked off. If you think holistically, it is not difficult to see how this can be proceduralized—software, extensions, services, marketing and so on.

As a further example of the granularity that organizations need to adhere to as digitization maturity embeds itself into operational cultures at unprecedented paces, the decomposition of the data taxonomy into further efficacy criteria shows impacts of scale, speed, prioritizations, and even the rules of engagement. Below are a few of the representations along the various taxonomies that demonstrate clear clusters of priorities when FFSBO's properly assess their digitization needs.

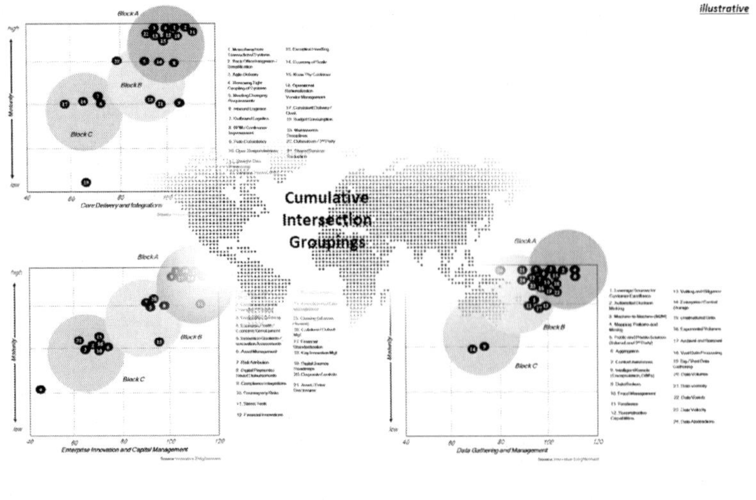

With the rebuilding of the financial markets, organizational leaders in FSBO's recognize that digitization has become so prevalent that the choices for assembling a sustainable solution set are overwhelming and the outcomes far too unpredictable. It's a reminder that to layer technology solutions without first implementing customer service and market facing functions (as examples) may result in worse financial performance than if nothing was undertaken.

Additionally, there needs to be more than a traditional sponsor of digitization within business units. Just as we are witnessing with its tangential cousin big data, all digitization efforts must be part of a larger roadmap (hence the 7,500+ intersection points). The justification for this resides in a very useful and proven set of guidance involving business orchestration as identified by widely published Agile intellectual capital.

A scant few examples of digitization business and operational misconceptions beyond simply choosing a sponsor include:

- Digitization is not a business model—it is just a means to an end, like architecture is to infrastructure.
- Mobile apps are not cash cows (or a business model)—only when they become a continual core for enlightened innovation (by destroying or expanding what is commonly available).
- Bespoke digitization's are the future to servicing investors and consumers.
- Digitization data is not big or small—it's really how it is insightfully used and protected along with quality and retirement.
- Digital banking is not about government—it is only if the system propagates century old axioms to modern, global realities of currencies, settlements, risk premiums, and discovery.
- Risks are controlled (and controllable) by regulators, solution sets, vendors and rules—ah, no.

Treating the perpetual demands of digitization as a

one-off project to implement and forget (the paradigm of a sponsor approach), fails to account for changing consumer behaviors, rapid cycle technologies (e.g., experimentation), and advancements in software and P2P media interactions.

In FSBO's, a few axioms for digitization program initiatives include:

- Stitching hardware, software and services layers requires forethought, afterthought, skills, and commitment.
- Connecting and automating digitized workflows requires a cradle-to-cradle culture.
- What is not always a formula for success is that, "if you build it they will come."
- "Me-too[ciii]" strategy and architecture adoption will yield, at best, commodity-like returns.
- Domino effects need to be anticipated and incorporated into larger programs.
- Data digitization is like an amorphic sand pile—one grain in a piece of a larger solution that until mixed with agents (i.e., applications), is still just sand.

This abbreviated list of axioms is reinforced by my research that show there are roughly 160 top priorities spread across "Block A" functional programs. Therefore, digitization priorities must fit a wide range of needs and in reality, only 8 to 12 per block totals (i.e., A, B or C) should ever be attempted in a 12-month rolling period.

Digital leadership materials point to a need for a new approaches and adaptive solutions executed against a framework for growth using stepwise improvements targeting a future state of operational excellence. But to soundly implement digitization solutions, leaders are forced to question assumptions and enterprise responses:

Are corporate leaders playing tit-for-tat with politicians all in an effort to outlast regulatory reforms thereby making the industry seem like a bunch of opportunists and knockers?

Along with repayment of past deeds (e.g., poor quality MBS's, origination practices and rampant greed), are enterprises continuing to ignore what it is they should be doing to achieve a liquid market beyond a common security (which itself brings a new set of untested uncertainties—it's not a panacea)?

Risk is being driven out of the systemic financial supply chain (e.g., prospecting, originations, servicing, instruments, secondary, investors), but is there a role for risky products which satisfy informed investor demands?

Is "group-think" at the hands of politicians and regulators creating another set of shortcomings all in the name of "safeness and soundness"?

Is living-and-dying by day-to-day operations (e.g., origination costs, loan modifications, product margins, technology) clouding new lending and banking models?

After nearly a decade since the start of the crisis[civ] began unfolding, what really has changed for the positive when examined from a consumer viewpoint (e.g., the industry says governance and oversight, but does the homeowner really see the desired benefits or are we taking regulatory communications at face value)?

All these questions, moving parts, and industry organizations apparently are assuming that regardless of the noise across the delivery channels they will incrementally adapt to keep pace with change. Achieving an end-state (e.g., strategic directions, implementing vision, developing safe and sound strategies), requires that we lift our heads up from the chaos and seek out directionality and proving conclusions, digitization as we practiced it for decades has died.

Moving forward, there will be issues with faster settlements, digital currencies, cyber protections (across every enterprise facet and systemic), disposable hyper-technology, narrow or disappearing channels, by-passing of systemic systems (e.g., settlements, transactions, credit reporting), and of course, new entrants that don't play by or are obligated to adhere to the "old rules, goals, and objectives."

Appendix B—The Rise of Big Data

(An abstraction from "**Big Data: Savior of Enterprises or a False Prophet?**"), written by Mark Dangelo and Cary Burch, 2015.

Data, data, data everywhere—but what to do with it when it increases by two to three times the vast stored amount already in existence every year? The logical expansion of data is moving from text to sound to video and pictures being most of the largest increases, with some estimates that in the next ten years, storage of bits and bytes will increase over 100-fold, pushing existing disciplines, analysis and usability techniques to near worthlessness.

Let's further examine the trends and figures:

- Using a compilation of projections, it is estimated that by the end of 2014 computer systems **had stored 2.2 x 10^{23} bytes of data**. Translated differently, this is the approximate equivalent of 20 billion terabytes or 4 terabytes for every person in the world. This estimation minimally **doubles (compounded) every year** with 2015 having 8 TB and 2020 having 22 TB per person respectfully.
- Per IDC, the number of **computer servers** needed to record and manipulate this vast amount of information will increase at least 10-fold by 2020 (this also considers doubling of computing power every 18-24 months which add tens of factors to existing systems)[1].
- Per IBM, 1.3 Exabyte's of information are sent and received by **mobile users** per year, 2.9 million **emails** are sent every second, each **household**

[1] "*Worlds data will grow 50x in next decade*", Computerworld, June 28, 2011.

consumes 375 megabytes each day, and 700 million minutes are spent on **Facebook** every month[2].

- IT spending breeched $1 trillion globally just five years ago and within this amount **data storage and management has been increasingly by 40% to 95%** depending upon industry and segment. With a shift to cloud solutions to deal with capital expense limitations, external the enterprise, hundreds of billions are being spent to deal with data growth but conversely, privacy, security and regulatory oversight have also escalated.
- **Key management disciplines and solutions** surrounding data security, encryption, risk management, archiving and destruction, reporting (e.g., consumer, regulatory, investor), and other co-dependent activities all surrounding the capture and keeping of data have also increased by factors and complexity.
- Huge changes in the **amounts and volumes of financial transactions** have pushed global debt to over $100 trillion[3], financial mobile usage has increased to nearly 50% of global smartphone users, and 4,000,000 Terabyte's per month (with a **compounded growth rate of 92%)[4]** are generated for data analysis.

In general, there are disciples and naysayers when it

---

[2] What is telling about these statistics, linking trends from within the IDC Digital Universe Study in 2011-2012 in conjunction with other sources, is that 75% of the data generated is by individuals, but 80% of this information is the responsibility of companies to protect, secure and store.

[3] This represents a nearly 40% increase since 2007 per **BIS**—Bank for International Settlements. See "*Debt Exceeds $100 Trillion as Governments Binge*", Bloomberg, March 10, 2014.

[4] Per Cisco Systems, with an estimate for 2017 at 23,000,000 terabytes month. Additionally, smartphone penetration rates are near 40% of the world's population with nearly 70% having some type of mobile device.

comes to trying to validate the aforementioned projections. Still, we cannot deny that with advancements in technology and increasingly sophisticated software, enterprises and consumers are able to deploy and consume data in volumes incomprehensible just a decade ago.

With ever more capable interconnections and the proliferation of millions of specialized apps, our enterprises and trading partners create and consume more data than believable[5].

So, as IT budgets rise sparingly, where are the skills and personnel needed to assess, secure, manage and extrapolate projections going to come from? Is the promise and lure of riches within these "big data" sources or "caves" leading to measurable profits, or will they fail to deliver once they have been explored[6]?

Beyond early limited adoption success or the transplanting of national security methods into the private sector, the enterprise use of big data has been one of caution—before conversion.

Consistently when we scan the media and columns headlines there seems to be some journalist engaged in promoting, discussing or condemning the use of data. Data has become the blood of economic engines, counter-terrorism, marketing, product development, cost containment, legal discovery, personal preferences, and even enlightenment.

Historically, data was encapsulated in department or corporate silos—locked up due to privacy, security and as a

---

[5] "A glut of available data is forcing (in-house) economists to serve as gatekeepers to ensure that disparate units within companies are using the same data sets and information inputs in their forecasting", "*In House Economists are Hot*", The Wall Street Journal, February 27, 2014.

[6] Internet searches (e.g., Google, Yahoo) do not constitute "big data"—that is information accessed by the consumer and in many case a creation source of big data. Big data in the context of this paper refers the enterprise usage as most big data efforts and their ensuing usage within functional data caverns are folding together vast amounts of consumer data along with other non-public private data to arrive at insights, forecasts, validations, and even innovations.

corporate competitive asset[7]. This data was not for sharing, let alone for collaborative extrapolation to predict bespoke markets and individual behaviors.

In simpler terms, the wholesale collection of data and its parsing and integration with external purchased data broker sources was not economically feasible a decade ago. As data storage devices became cheaper, larger and even virtual (i.e., cloud arrays), storage of once transitory transactions are held for extended periods of time swelling repositories and creating cross-industry estimated revenues over $25 billion[8] for 2014 in the process just to mine and access these bits and bytes.

Within the same time, the idea of "big" or "vast" data emerged as stored data extended beyond 20,000 billion terabytes of data and began doubling every year. By linking together these data sources using technically advanced solutions (e.g., massively parallel processing, grid computing), the broad goals were to:

- **Ensnare and serve customers with excellence[9]** when and where they wanted corporate products and services. By understanding the minutia of consumer behavior, offerings, markets and delivery could be mass produced, but customized discretely to buyers without extensive capital improvements.

---

[7] An interesting quandary has arisen—if the consumer provides a majority of data in public forums, is it "private" and how should it be protected? With mobile communication volumes already in the Exabyte's and doubling every two years, is using a defense of "...(we've) generally gone farther than most apps to protect data" a sufficient defense should a breach of trust or systems occur? See, "*PSST, Your Online Secret Isn't So Secret*", The Wall Street Journal, February 26, 2014.

[8] Current projections place the growth of big data to be over 30% compounded yearly subject to continual upward revisions as services push the adoption of more complex services, hardware and software. It should also be mentioned that some estimates place big data at over $50 billion in total market size.

[9] Early references to massive CRM solutions were also focused on customer excellence—nearly two decades old.

- **To gain competitive advantages**[10] to not only retain customers, but to expand market share using predictive capabilities and bespoke arrangements. Moving beyond a "one-size-fits-all" method, incorporating mobile and segmented delivery technologies in partnership with telecommunication providers, operational delivery can be expanded in concert with operational efficiencies.
- **Hone business processes**[11] by maximizing customer and channel demands, while phasing out or eliminating those segments that failed to meet returns using segmented data building blocks. By encapsulating data functions, new solutions could be built quicker, with greater success and with less risk.
- **With consumer activities**[12] radically altered during the last economic recession, development of new profiles and models must be proven to ensure profitability. As consumers become more self-reliant and able to research product options with their own data sources, traditional marketing approaches must evolve to reach social media, mobile and otherwise content-hungry data users.
- **Predict behaviors**[13] that include consumer profitability, likelihood of financial default, buying

---

[10] A key goal of the business intelligence (BI) community when it first arrived in business units and much has been written and developed in terms of methods and techniques to aid with this goal.

[11] This was a goal within the BPM and process reengineering efforts to aid with the streamlining of efforts and to improve operational efficiencies.

[12] Most developed nations were consumers of goods and carried heavy personal debt burdens. When buying habits changed in such a large scale and confidence is lost.

[13] Dealing with the Great Recession of 2007-2011, firms quickly realized that their models for predicting consumer reactions were inaccurate and failed to anticipate occurrences such as "strategic defaults".

patterns, household profile, risk archetype, and many more industry benchmarks and discrete dashboard measurements.

However, what all these data goals addressed using big, vast, small, medium, local, remote, summarized, discrete, farmed, extrapolated, and many other collection forms and variants is that until recently, their deployment was about efficiency and product adoption—if it could be analyzed timely and with sound interpretation[14]. What has become progressively adopted by executives and boardrooms is that the unlocking of these far-flung data repositories and discrete caches can uncover M&A opportunities, hidden innovations, and even unconventional financing solutions.

Data has become progressively more about expanding the topline rather than just controlling the bottom line for enterprise profitability. However, what remains to be seen is whether ***big data can be a sustainable savior*** to aid management teams to survive and prosper (e.g., top line growth), or whether it will become so unwieldy and complex that the faith placed in it and the money spent to acquire it make the faith in data ***a false prophet that condemns its supporters to ruin***.

To answer the question—savior or false prophet—against the backdrop of deployment expectations and statistics, we need a methodical examination of how to seamlessly incorporate big data and its "caverns" across organizational culture, technologies, strategic intent, security and privacy and even determination of success.

No two operations are alike and big data is currently not an off-the-shelf solution. Therefore, the remainder of

[14] "Data analysis is considerably more challenging than simply locating, identifying, understanding, and citing data. For effective large-scale analysis, all of this has to happen in a completely automated manner. This requires differences in data structure and semantics to be expressed in forms that are computer understandable, and then 'robotically' resolvable." **Challenges and Opportunities with Big Data**, Purdue University.

this prepared paper is broken into five segments[15]:

1. *What constitutes a "data cave" and what are its key characteristics?*
2. *How do organizations prepare for and adopt big data and its usage of caverns?*
3. *What critical components and architectures must be in place to not only leverage big data, but create the caverns?*
4. *What are the impacts on processes associated with technology integrations?*
5. *What are the skills and services that will likely be demanded to ensure initial success and on-going internalization?*

Contemplating the rapid advancement of big data solutions can be boggling to the mind—let alone deciding which one makes sense for the enterprise. As industries stretch the minds and imagination of leadership seeking answers, tiers and clusters of storage are emerging. Yet, for those FSBO's and their leaders seeking sustainable innovation, what are the options, who are the players, and what disciplines are solidifying?

Like most evolving innovative ideas, there is a lack of consistency in, "What constitutes big data?" With the revival of the in-house economist[16] and a desire to create an enterprise view of department and divisional data beyond blended dashboards, strategic leaders are shifting the data from local to holistic caverns, but using additional sources that are extracted from trading partners, data brokers and

---

[15] We use five sections here to demonstrate where big data actions fail to deliver against perceptions and expectations. Moreover, when reviewing the history and maturing technologies of big data from their early origins within business intelligence and data warehousing, it is clear that without strong and measured ties to business strategy, organizational efficiencies and growth projections, big data can become a disillusionment for leaders thereby slowing or cutting off funding past initial "failures."

[16] "*In House Economist are Hot Again*", The Wall Street Journal, February 27, 2014.

public information sites[17].

Today, nearly 50% of big data market revenue is achieved from professional services—companies hiring experts to aid with the definition and deployment of specialized big data hardware and software[18].

Critical changes in disciplines, trends and technologies necessary to achieve value-added big data solutions within the FSBO include:

- Hadoop has become the foundation for big data integration[19]. With an open-source architecture, the robustness and capabilities are growing at rapid rates creating repositories of solutions spanning both industry and functionality.
- Access to data using interpretative or visual programming languages from vendors is providing greater levels of abstraction delivering robust capabilities closer to the business leaders seeking answers.
- In-memory storage and analytics advance with capabilities doubling nearly every 12 months.
- The four V's—**volume, velocity, variety, and veracity**—are rapidly morphing with the solutions available as once emerging technologies and directions become hardy with new entrants taking their place in the target zones.
    - Volume: greater specializations, more segmentation, wide variations, privacy and governance changes.
    - Variety: dimensional facets, divergence and reconciliations.

[17] **IT Innovation Spurs Renewed Growth**, A.T. Kearney, 2013.

[18] **Big Data Vendor Revenue and Market Forecast 2012-2017**, Wikibon.

[19] "Top Technology Trends in Big Data", Lance Riedel, 2013

- Velocity: zero-time analysis, transitory predictions, vast sources.
- Veracity: Bias, algorithms, analytics, models, scalability all playing catchup.

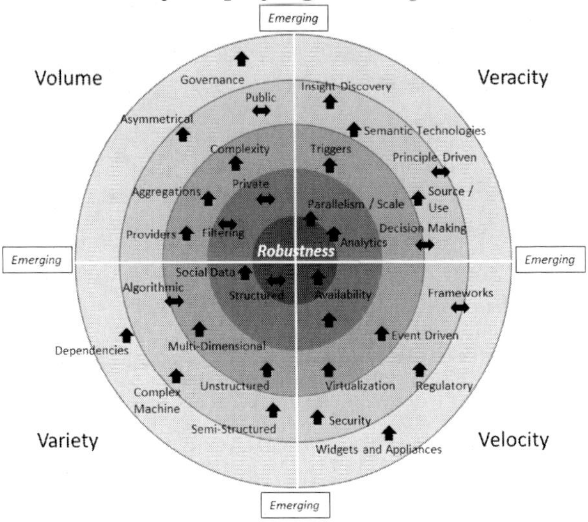

- The incorporation of semantic data (as defined by semantic technologies separating data, meanings, relationships and code) is seeking to model more real-world outcomes from the vast sources available to Chief Data Officers (CDO's) and their growing staffs of data scientists. It incorporates the "orchestration" of components in new and unique ways, while reducing delivery (capital) and maintenance costs to promote adaptability and improve execution times for decision makers.
- Infrastructural shifts are being deployed both within the enterprise and across virtual provisioned solutions[20] (e.g., infrastructure as a service—IaaS, platform as a service—PaaS). A few of these areas include: storage infrastructure, scripting and development tools, high capacity warehouses and

[20] "Analytics: The real-world use of big data in financial services", IBM Institute for Business Value, page 5

caverns, analytic accelerators, columnar databases and stream-based computing.

- As mentioned, big data sources supported by hundreds of thousands of consumer and industry widgets, are changing not only the volume and variety, but the velocity and veracity of their use. Sources[21] now include not just web pages and social media long sought after by marketers, but increasingly the use of:
    - Event data
    - RFID sources
    - Machine sensors
    - Global positioning
    - Free-form text
    - Images
    - Video
    - Audio
    - Transitory or temporary
    - "Phone home" devices
    - Retail
    - Public databases
    - Games
    - Entertainment
    - Businesses

The changes within technology are expected, but to fully internalize the challenges with big data, enterprises will need to examine fit and soundness before success can be achieved.

The Use of Data Caves

There is a historical parable that is found in many spiritual dogmas where a man enters and cave to find wonders and experiences divine guidance from the messages received from within. For enterprises and

[21] Ibid, page 6

governments storing and processing petabytes of data to arrive at conclusions, their drive for answers and enlightenment from the seemingly random collection of transactions, consumer postings, structured and unstructured elements, third-party data sources, outsourced data processors / collectors and other collective "data caves"[22] remains challenging and costly.

Like a spelunker twisting and turning to reach hidden or seldom seen portions of a cave system to witness a wonder of nature, data cavern advocates use these techniques to provide the boundaries and encapsulation, while ensuring that their iterative efforts can deliver results within set timeframes and with predictable results. However, depending upon what is being sought and how the pilgrim seeks to find answers, these same caverns, like the parables of ancient times, will yield varied actions for enterprises and their advocates.

So, if you have expectations or are seeking definitive answers from the caverns, these contrasting data sources framed within the confines of a cavern will reveal answers that align with what is being sought. Nevertheless, the danger of analyzing vast amounts of data from multiple points of origination and manipulation resides with arriving at projections or conclusions which are difficult to play "devil's advocate" against.

There is inherent danger in arriving at "answers" because that is "what we wanted to find." These "false positives" or even "false negatives" are increasingly and casually taken as gospel without due diligence resulting in career changing initiatives and organizational expense. In the end, the efforts and costs taken to arrive at the initial

[22] Data caverns are defined as huge collections of data elements, files, types and taxonomies brought together to form big data initiatives (within reusable and scalable maps that provide building blocks as normal parts of delivery). Like caves, a collection of big data is not steady state as changes and environmental factors conspire to alter predefined routes or patterns. Big data cannot efficiently exist without a data cavern—linked or mined repositories—to provide the means to arrive at conclusions and not result in "boiling the oceans" of data.

projections are not reciprocally available to "prove" that the big data found within the data caverns says something else entirely. Since big data and data caverns are enterprise initiatives—not consumer—these early efforts can either be used as leverage or as a stone to punish.

Therefore, enterprises need to guard against many actions like the pilgrim entering a cavern system for the first time seeking enlightenment[23].

- **Composites**: Data caverns **represent composites of components** and can be subject to erroneous conclusions at the hands of personnel who are inexperienced or improperly skilled at the sourcing and manipulation of disparate sources. For example, how many interpretations can be taken from the U.S. tax databases housed within the IRS systems alone? Imagine if the aforementioned were linked to other cavern systems (linkages of caverns and external big data sources) such as sovereign intelligence (e.g., NSA), healthcare systems, financial transactions and even Alibaba, Amazon or eBay supported by mobile telecommunication providers. How many answers could be returned when linking so many sources together? How many false-positives or false-negatives would be experienced?
- **Evolutionary**: Use of data caverns and big data in general is not a "one-off" designed to achieve a result and then be tossed away. By its very nature, **big data is meant to be an evolutionary process** for the life of the enterprise getting stronger and more insightful with age. Early on, big data initiatives were described to be more competitive, to predict changing consumer behaviors, to identify market opportunities or risks that others have yet to experience or even contemplate. However, as these

[23] Big data analysis without the utilization of a framed cavern or loosely coupled approach materially contributes to scope creep, poor deliverables, operational and process mismatches, and a lack of reusability and leverage.

early successes or even failures taught the organizations, there is more needed than just the will to create or secure the data itself. The enterprise requires preparation, understanding, a need, and of course, the means to accomplish measurable results rather than "cloudy" promises.

- **Expectations**: Before incurring time and expending treasure, **are we chasing a faith** or a vetted expectation? When we create our data cavern in the hopes of realizing the potential of advanced technology solutions (e.g., multi-dimensional mining, pattern recognition, massively parallel hardware and application threads), are we trying to build the solution to match capabilities or are we objectively compartmentalizing the strategies, architectures, measurements, equations, and relationships as part of an iterative framework, which leads us to a definitive roadmap? Success in harnessing the data cavern(s) is given to our operations as a divine twist of fate.
- **Layers**: Data caverns are evolving into (once each is complete with its initial "excavation", mapping and coupling) **complexes of caverns** each signifying progress and insight—interconnected layers of data if you will. They represent leading innovation, but as we have learned that while these caverns represent cutting innovation, they are also not random[24]. Caverns without concrete ties to business strategy[25]

[24] "...nearly 60 percent of respondents say they should invest in IT innovation at the leading edge and early adopter stages, yet only 27% indicate that they actually do." **IT Innovation Spurs Renewed Growth**, A.T. Kearney, 2013, page 3.

[25] What is oxymoronic is that there are big data advocates who believe the use of strategy to promote and adopt lasting solutions somehow stifles innovation. Innovation is not about random chance nor it is just luck nor is it "give me a bag of money and trust me on the results". Innovation by nature of its implicit definition is strategy tied to results and measures. Otherwise, it would be research and development (R&D). Big data and caverns require strategy to align results and to marshal the organization (and its culture) behind its internalization. Without strategic alignment, big data and their caverns become one-offs and expensive white elephants.

and measurements will not likely survive budget reviews and without measurements provided by direct projects or programs, the value and returns promised cannot be achieved let alone validated.

Caverns of big data sources are in their infancy. As a result, they are promising, but prone to unrealistic expectations. Conversely, use of big data deployed alongside sound and vetted business achievements can represent huge enterprise and market payoffs that have not been witnessed since the early 1990's with the introduction of data warehousing solutions.

Appendix C—Acronym and Phrase List

4V's – Volume, Velocity, Variety, Veracity
ABA – American Bankers Association
AI – Artificial Intelligence
AOL – America On Line
API – Application Programming Interface
ATM – Automated Teller Machine
AVM – Automated Valuation Model
B&M – Brick and Mortar
B2B – Business to Business
B2C – Business to Consumer
B2G – Business to Group
BaaP – Banking as a Platform
BaaS – Banking as a Service
BAU – Business As Usual
Big-5 – JP Morgan Chase & Co., Bank of America Corp., Wells Fargo & Co., Citigroup Inc., and Goldman Sachs Group Inc.
BP – Business Process
BPO – Business Process Outsourcing
C2C – Consumer to Consumer
CAGR – Compound Annual Growth Rate
CCMR – Committee on Capital Market Regulation
CD – Certificate of Deposit
CDO – Chief Data Officer
CEO – Chief Executive Officer
CFO – Chief Financial Officer
CFPB – Consumer Financial Protection Bureau
CIO – Chief Information Officer

CMO – Chief Marketing Officer
COE – Centers of Excellence

CRO – Chief Risk Officer
CSC – Computer Sciences Corporation
CTO – Chief Technology Officer
DDA – Demand Deposit Account
DLT – Distributed Ledger Technology
DOE – Digital of Everything
EC – Electronic Currencies
EU – European Union
FDIC – Federal Deposit Insurance Corporation
FED – U.S. Federal Reserve
FinTech – Financial Technology
FSBO – Financial Services and Banking Organization
Gen X – Generation X
GDP – Gross Domestic Product
GSE – Government Sponsored Enterprise
HELOC – Home Equity Line of Credit
HR – Human Resources
HUD – Housing and Urban Development
IaaS – Infrastructure as a Service
IC – Intellectual Capital
ICO – Initial Coin Offering
IoT – Internet of Things
IT – Information Technology
ITO – Information Technology Outsourcing
IV&V – Independent Verification and Validation
JV – Joint Venture
KYC – Know Your Customer
M&A – Mergers and Acquisitions
MBA – Mortgage Bankers Association
MBS – Mortgage Backed Security
OCC – Office of Comptroller of Currency
PaaS – Platform as a Service
P2P – Person to Person or Peer-to-Peer
PU – Profit Underruns
R&D – Research and Development

RFID – Radio Frequency Identification
RI – Responsible Innovation
S&L – Savings and Loan
SEC – Securities and Exchange Commission
SOP – Standard Operating Procedure
T+0 – Trade plus zero days
TARP – Troubled Asset Relief Program
TBTF – Too Big To Fail
TBTS – Too Big to Succeed
TDA – Term Deposit Account
TRID -- TILA-RESPA Integrated Disclosure
VC – Venture Capitalist
VR – Virtual Reality
VRU – Voice Response Unit
WEF – World Economic Forum
WSJ – Wall Street Journal

Acknowledgments

Special thanks go out to those professionals who contributed their time and ideas to help this book reach its conclusion. Thanks go out to Pedro Fong for his input and guidance along with some meaningful changes that resulted in a title change for this book.

Of course, thanks go out to our families for putting up with our constant work over the last year on this book and the struggles to push it to its conclusion.

Bibliography / References

"*Accelerating the Digitization of Business Processes*", McKinsey.com, 2015.

Almquist, Eric, Mitchell Leiman, Darrell Rigby and Alex Roth, "*Taking the measure of your innovation performance*", Bain & Company.

"*Analytics Trends 2015 A Below-the-Surface Look*", Deloitte, 2015.

Armstrong, Paul, "*Three Big Trends in FinTech You Should be Using in your Business Today*", Forbes, June 27, 2016.

"*Backing up the Digital Front: Digitizing the Banking Back Office*", Capgemini Consulting, 2013.

Baculard, Laurent-Pierre, Greg Caimi, John Senior and Elizabeth Spaulding, "*Leading a 360-Degree Digital Transition*", Bain & Company.

Baert, Rick, "*11 banks complete experiment using blockchain technology*", Pensions & Investments, January 22, 2016.

"*Banking in India Downwardly Mobile*", The Economist, February 15, 2015.

Baxter, Mike, Steve Berez and Vishy Padmanabhan, "*Rebooting IT: Why Financial Institutions Need a New Technology Model*", Bain Brief, September 4, 2014.

"*Big Banks Following FinTech Startup Trends*", Global Banking & Finance Review, January 13, 2016.

Black, Jeff, John Follain, "*Davos Elite Seeks Fixes to Defend the System From Populists*", Bloomberg, January 18, 2017.

Bradley, Chris, Lowell Bryan and Sven Smit, "*Managing the strategy journey*", McKinsey Quarterly, July 2012.

Brear, David and Pascal Bouvier, "*Exploring Banking as a Platform (BaaP) Model*", The Financial Brand, March 4, 2016.

Brito, Michael, "*What you Thought about B2B Marketing and Millennials Might be Wrong!*", LinkedIn, March 17, 2015.

Broeders, Henk and Somesh Khanna, "*Strategic choices for banks in the digital age*", Mckinsey & Company, January 2015.

Brunner, Simon, "*Digitalization: Banks are at a Crossroads*", Credit Suisse, 2015.

Burns, Matt, "*This Could Be the Mortgage Industry's iPhone Moment*", Tech Crunch, November 24, 2015.

Carlson, David Gray, "*The Securitization of Mortgages in the United States: A Market of Mass Destruction*", Benjamin N. Cardozo School of Law.

Chavez-Dreyfuss, Gertrude, "*Digital currency sales take off, but with no regulation questions abound*", Reuters, December 19, 2016.

Chinn, David, James Kaplan, and Allen Weinberg, "*Risk*

and responsibility in a hyperconnected world: Implications for enterprises", McKinsey & Company, January 2014.

"Cloud Contact Center Total Cost of Ownership" infographic, TeleTech, November 9, 2015.

Cognizant, "*The Networked Bank: Utility Concepts to Transform the Operating Model*", 20-20 Insights, September 2014.

Cognizant, "*How Banking as a Service Will Keep Banks Digitally Relevant and Growing*", Keep Challenging, June 2016.

"*Common Securitization Platform and Single Security*", Booz Allen Hamilton, 2015.

"*Continuous Delivery: A Maturity Assessment Model*", Forrester Consulting, March 2013.

"*Controlled and Managed Transformation Unlocking Innovation with Data and Analytics*", HCL Technologies, 2016.

"*Corporate Learning 'Wildly Out-of-Sync' with How Individuals Learn for Themselves*", Towards Maturity, February 2, 2016.

Court, David, "*Getting big impact from big data*", Mckinsey Quarterly, January 2015.

Couto, Vinay, John Plansky and Deniz Caglar, Fit for Growth, John Wiley & Sons, Inc., Copyright 2017, Pricewaterhouse Coopers Advisory Services, LLC.

"*Covered bonds back where they belong, but LCR impact in doubt*", The Cover, January 8, 2015.

Coward, Jeremy, "*Outsourcing in 2020*", Outsourcing Yearbook 2016, 2016.

"*Creating a Competitive Advantage*", TD Bank.

"*Crisis management Focus on: Building crisis-ready boards*", Deloitte, May 2015.

Crosman, Penny, "*Fed Hires IBM to Modernize ACH Amid Push for Faster Payments*", American Banker, January 13, 2016.

"*Cyber Crime: The Fast-Moving Menace—A Special Report*", Thomson Reuters.

Dangelo, Mark P., "*Abundant Innovative Disruption—Varian's Rule Is Almost Upon Us*", Mortgage Bankers Association, March 2016.

Dangelo, Mark P., "*Banking Supply Chains Are Not Just About Customers*", Mortgage Bankers Association, April 2017.

Dangelo, Mark P., "*Banking Utilities versus Banking Innovators—Which Will You Become?*", Mortgage Bankers Association, August 2016.

Dangelo, Mark P., "*Being Cool, Being Trendy, Being Hip—Rethinking FSBO*", Mortgage Bankers Association, June 2016.

Dangelo, Mark P., *Beyond Corporate Project Mentality*, January 2016.

Dangelo, Mark P., "*Beyond the Headlines—Using IT Substitution Across Functions*", Mortgage Bankers Association, October 2016.

Dangelo, Mark P., "*Chasing the Digitally Driven—Part 1*",

Mortgage Bankers Association, July 2015.

Dangelo, Mark P., "*Chasing the Digitally Driven—Part 2*", Mortgage Bankers Association, August 2015.

Dangelo, Mark P., "*Chasing the Digitally Driven—Part 3*", Mortgage Bankers Association, October 2015.

Dangelo, Mark P., "*Chasing the Digitally Driven—Part 4*", Mortgage Bankers Association, November 2015.

Dangelo, Mark P., "*Chasing the Digitally Driven—Who's Conducting the Transformation?*", Mortgage Bankers Association, December 2015.

Dangelo, Mark P., "*Currencies—The Next Quantum Advancement in Digitization*", Mortgage Bankers Association, January 2016.

Dangelo, Mark P., "*Deregulation: An Innovation Damper for a New Year?*", Mortgage Bankers Association, January 2017.

Dangelo, Mark P., "*FSBO Responsible Innovation—Consumer Protection or Regulatory Nightmare?*", Mortgage Bankers Association, May 2016.

Dangelo, Mark P., "*Outsourcing in the Next Presidential Cycle*", Mortgage Bankers Association, September 2016.

Dangelo, Mark P., "*Regulatory Easement, Insulating Realities*", Mortgage Bankers Association, January 2017.

Dangelo, Mark P., "*Rhetoric will Require a Shift in Employment Strategies*", Mortgage Bankers Association, March 2017.

Dangelo, Mark P., "*Technology, Folly in an Age of Commoditization?*", Mortgage Bankers Association, February 2017.

Dangelo, Mark P., "*The Game of Maneuvers—Strategy's 'Game of Thrones'*", Mortgage Bankers Association, May 2016.

Dangelo, Mark P., "*Tunnel Vision—Competitively Doing What Others Do for the Wrong Reasons*", Mortgage Bankers Association, March 2016.

Dangelo, Mark P. and Cary Burch, *Big Data: The Savior of Enterprises or a False Prophet?*, May 2014.

"*Data Access Governance*", EMC.

"*Deloitte's 2016 Global Outsourcing Survey*", Deloitte, May 2016.

"*Developments in financial services sector requires new skills*", PwC Actueel.

"*Digital Business Era: Stretch Your Boundaries*", Accenture, 2015.

Dilts, Elizabeth, "*Third of consumers open to Goggle, Amazon banking: Survey*", Technology News, January 11, 2017.

Donnan, Shawn, "*Air Pollution Deaths Cost Global Economy $5TN Annually*", Financial Times, September 8, 2016.

"*Driver Debt*" graphic, Bloomberg, February 17, 2017.

Duncan, Ewan, Eric Hazan and Kevin Roche, "*Digital disruption: Six consumer trends and what businesses need to do now*", McKinsey, March 2014.

Eaton, Ross and Peter Reynolds, "*Ring-Fenced 14 Success Factors for IHC Board Governance*", Oliver Wyman, July 2016.

Ebert, Michael, Mark Johnson and Bill Dieringer, "*Managing the Risk: Recent Data Breaches Bring Security up the Healthcare Agenda*", KPMG, 2015.

Elumalai, Arul, Irana Starikova and Sid Tandon, "*IT as a service: From build to consume*", McKinsey & Company, September 2016.

"*Embracing technology for rapid transformation—data, digitization and disruption*", KPMG Capital Markets, 2015.

Evans, Phillip, and Patrick Forth, "*Navigating a World of Digital Disruption*", BCG.Perspectives, April 29, 2015.

FDIC Quarterly, Federal Deposit Insurance Corporation, Volume 9, Number 1, 2015.

FDIC Quarterly, Federal Deposit Insurance Corporation, Volume 9, Number 2, 2015.

FDIC Quarterly, Federal Deposit Insurance Corporation, Volume 9, Number 3, 2015.

FDIC Quarterly, Federal Deposit Insurance Corporation, Volume 9, Number 4, 2015.

FDIC Quarterly, Federal Deposit Insurance Corporation, Volume 10, Number 1, 2016.

FDIC Quarterly, Federal Deposit Insurance Corporation, Volume 10, Number 2, 2016.

FDIC Quarterly, Federal Deposit Insurance Corporation, Volume 10, Number 3, 2016.

FDIC Quarterly, Federal Deposit Insurance Corporation, Volume 10, Number 4, 2016.

FHFA, "*Building a New Infrastructure for the Secondary Mortgage Market*", Industry comment, October 4, 2012.

"*Financial Supply Chain Management*", Deloitte, 2013.

Fratantoni, Michael, "*Who Will Own Mortgage Assets?*", Mortgage Bankers Association, 2014.

Gandhi, Niyum and Tomas Mikuckis, "*Getting Ready for Price Transparency*", Oliver Wyman, 2014.

"*Going Digital: The Banking Transformation Road Map*", A.T. Kearney, 2014.

Groenfeldt, Tom, "*Moven Takes It Mobile Banking Global with Accenture*", Forbes, May 13, 2015.

"*GSE Reform Principles and Guardrails*", Mortgage Bankers Association.

Henry, David, "*U.S. bank group to launch Zelle instant payments brand in 2017*", Reuters, August 26, 2016.

Hirt, Martin, and Paul Willmott, "*Strategic principles for competing in the digital age*", McKinsey Quarterly, May 2014.

"*How Low Could We Go? Negative Interest Rates Explained*", Charles Schwab, April 13, 2016.

"*How outsourcing relationships will change towards 2020*", ServiceFutures.com, June 13, 2016.

Hudson, Kris, "*Why a Stronger Housing Sector Isn't Boosting the U.S. Economy That Much*", The Wall Street Journal, September 9, 2015.

"*Hyperconverged infrastructure adoption is accelerating*", Helpnetsecurity.com, June 20, 2016.

"*Inside the Mind of Millennials: What Financial Marketers Need to Know*", The Financial Brand, January 25, 2016.

Irrera, Anna, "*'Regtech' startups see more business in Trump Era*", Technology News, December 22, 2016.

"*IT 2020: Preparing for the Future*", A.T. Kearney, 2013.

Jeans, Neil, "*Risk-Based Approach to KYC: Sound Concept, Complex Reality*", Thomson Reuters, 2016.

Jenkins, Patrick, and Martin Arnold, "*Beyond Banking: under attack on all sides*", Financial Times, November 10, 2015.

Kaplan, Robert S, and David P. Norton, "*Using the Balanced Scorecard as a Strategic Management System*", Harvard Business Review, July-August 2007.

Karabell, Zachary, "*The Uberization of Money*", The Wall Street Journal, November 6, 2015.

Kashani, Kamran, "Fight*ing Commoditization Strategies for Creating Novel Customer Values*", IMD.ch, October 2006.

Kaufeldt, Haley, "*Digitization of Business through Co-Creation*", Symbio.

Killian, Christopher B and Joseph Cox, "*The U.S. Private Label Mortgage-Backed Securities Market in 2015*", Sifma, July 1, 2015.

Kusisto, Laura, "*Renters Will Continue to Struggle for the Next Decade, Harvard Study Says*", The Wall Street Journal, 2015.

Kusisto, Laura, "*The U.S. Homeownership Rate Falls Again, Nearing a 48-Year Low*", The Wall Street Journal, April 28, 2016.

Lalic, Zoran, "*Risk Management: Risks are Lurking Everywhere*", Insecuremag.com,

Leondis, Alexis, "*Why Lenders Love the $2.5 Million Home Loan*", Bloomberg Business, February 24, 2015.

Long, Heather, "*Be 'very afraid' about globalization's next phase*", CNN, December 23, 2016.

Marko, Kurt, "*Hyperconverged infrastructure will evolve into cloud building blocks*", Diginomica.com, May 16, 2016.

Marous, Jim, "*10 Marketing Trends the Banking Industry Can't Ignore*", The Financial Brand, February 29, 2016.

Marous, Jim, "*Top 10 Banking Trends for 2016*", The Financial Brand, January 15, 2016.

McBride, Sarah, "*U.S. VC Firms Have Raised $32 Billion So Far This Year*", Bloomberg, October 10, 2016.

Mee, Paul, Kenan Rodrigues, Hilda Tang and Vanni Parmeggiani, "*Imagine if Your Data Were Good*", Oliver Wyman, 2015.

Meisler, Laurie, Mira Rojanasakul and Jeremy Scott Diamond, "*Who Gets Venture Capital Funding?*", Bloomberg, May

25, 2015.

"*Merchant Consultation Program Summary Report 2015*", Cash Product Office, 2015.

Mikkelsen, Randall, "*Trump's win clears way for reshaping financial regulation and compliance*", Thomson Reuters Blog, November 11, 2016.

"*Millennial Banking Insights and Opportunities*", FICO, 2014.

"*Millennials and Banking*" infographic, American Bankers Association, aba.com/Millennials.

"*Mobile Self-Service Redefines the Banking Experience*" infographic, IDC, Novemberr 2015.

"*Mobility and the State of Customer Loyalty in Banking*" infographic, Bain.com, November 2015.

"*Monetary Policy Normalization—Understanding the Tools Used by the Federal Reserve*", Federal Reserve, January 13, 2016.

"*North America Mortgage Banking 2020*", Accenture.

Nasiripour, Shahien, "*Student Debt in America Has Hit a New Record*", Bloomberg, February 2, 2017.

Pilon, Mary, "*Why Single Women are Buying Homes at Twice the Rate of Single Men*", Bloomberg, January 31, 2017.

"*Protecting Your Data Wherever It Goes*", Seclore, Infographic.

Reddy, Hema, "*Big Data Trends: The Top Eight Analytics Lessons for Business*", IBM, September 21, 2015.

"Retail Banking 2020 Evolution or Revolution?", PwC, 2014.

Ritholtz, Barry, "*Maybe We Clamped Down Too Hard on Finance*", Bloomberg, August 26, 2016.

Robinson, Edward, "*Mac McQuown is Reinventing the Corporate Bond*", Bloomberg Business, March 2015.

Rogers, Dale, Rudolf Leuschner and Thomas Y. Choi, "*The Rise of FinTech in Supply Chains*", Harvard Business Review, June 22, 2016.

Sarris, Bill, "*Escaping the Doghouse: Winning in Commoditized Markets*", BCG.Perspectives, 2016.

Schmidt, Lars, "*Why Retention Will be the Biggest Talent Challenge of 2017*", Forbes, December 16, 2016.

Schroeder, Pete, "*Banks spent record amounts on Lobbying in recent election*", Reuters, March 8, 2017.

Schwab, Klaus, The Fourth Industrial Revolution, Crown Business, Copyright 2016, World Economic Forum.

Sen, Conor, "*Conditions are Ripe for a Big-City Exodus*", Bloomberg, September 9, 2016.

Simon, Ruth, "*Tech Firms Venture into New Territory: Lending*", The Wall Street Journal, September 16, 2015.

Sinnock, Bonnie, "*Secondary Market Seeks Balance Between Private and Public Capital*", National Mortgage News, September 6, 2016.

"*6 Drivers for Moving Business to the Cloud*", Information Age, November 11, 2016.

Smith, Jeremy, "*IT Outsourcing 2020*", Maturity.com, 2014.

Sorensen, Bard, "*Schwab Sector Views: A Housing Update—and Why it Matters*", Charles Schwab, April 28, 2016.

Spaulding, Elizabeth, Sarah Dey Burton and Greg Caimi, "*Anatomy of a digital forerunner*", Bain & Company.

"*State of Regulatory Reform 2017: A Special Report*", Thomson Reuters, 2017.

Stewart, Luke A., "*The Impact of Regulation on Innovation in the United States: A Cross-Industry Literature Review*", Information Technology & Innovation Foundation, June 2010.

Stumbles, Peter, Mohit Wadan Andrew Carleton and Jason Heinrich, "*Big, Bold Bank Cost Transformation*", Bain & Company.

Tassabehji, Rana, "*The rise of biometric banking as fight against fraud is stepped up*", The Conversation, September 9, 2014.

"*Ten key points from Donald Trump's electoral victory*", PwC, November 17, 2016.

"*The Administrative Procedure Act and Federal Reserve Stress Tests*", Committee on Capital Markets Regulation. September 2016.

"*The Innovation Death Spiral*", Accenture, 2011.

"*The Intelligent Distributor*", Oliver Wynam, 2015.

"*The Power to Perform: Human capital 2020 and beyond*", PwC, 2016.

"The Rise of the Digital Bank", McKinsey.com, 2015.

"*The Rise of the New Marketing Organizations Data-Driven and Digitally Savvy*", Forbes, e.turn.com.

"*The Role of the Secondary Market in Mortgage Financing*", BiPartisanpolicy.org.

"*The Strategic CIO*", The Economist, March 2015.

"*The Theory of How the Financial System Created AAA-Rated Assets Out of Subprime Mortgages*", IMF, Global Financial Stability Report, April 2008.

"*Trump Cites Friends to Say Banks Aren't Making Loans. They Are*", Bloomberg, February 2, 2017.

"*Top Regulatory Trends for 2016 in Banking*", Deloitte, 2016.

Torres, Craig, "*As Wall Street Thrives, America's Little Guy Chokes on Paperwork*", Bloomberg, February 2, 2017.

U.S. Department of the Treasury, "*Opportunities and Challenges in Online Marketplace Lending*", May 10, 2016.

Verma, Sid, "*A Cheat Sheet on the Deglobalization of the Financial World*", Bloomberg, December 6, 2016.

Whalen, Christopher, "*Why Nonbank Lenders Are the Future of Mortgages*", National Mortgage News, February 16, 2016.

"*Wipro Launches Open Banking API Platform*", Wipro,

November 7, 2016.

"*What's Compliance Worth?*", Thomson Reuters, 2015.

Yurcan, Bryan, "*'Banking as a Service' for FinTech's Seeking Scale*", American Banker, March 21, 2016.

Zandi, Mark, "*The Biggest Challenge Facing Businesses in 2017*", Fortune, December 26, 2016.

Index

Author's Biographies

Mark P. Dangelo
Email:
mdangelo@mpdl2c.com
LinkedIn: mpdangelo

Mark P. Dangelo is a problem solver with an aptitude for identifying causality and actions. As an advisor, scientist, and innovator, Mark has worked domestically and internationally assisting hundreds of clients understand and profit from emerging operations, technologies, and workforces. He has advised numerous Fortune 500 firms and their executives on post-deal M&A integration efforts, new markets, operational efficiencies, and organizational restructurings.

Mark's line and c-level experience has been created from a hands-on foundation of business and technology expertise and delivery. His thought leadership, expertise, and corporate brand are recognizable within several industries due to extensive publications, articles, conferences, and books.

Mark works independently as a SME with a host of international individuals and organizations helping teams understand the relevant application of innovative ideas. He earned a Bachelor's of Science in Computer Science and a Master's in Business.

Rick Grant
Twitter: @nyrickgrant
Email: rick.grant@rga-pr.com.

Rick Grant owns RGA Public Relations, Inc., Oregon, Wisconsin, a communications firm that specializes in promoting companies operating in the U.S. home finance industry. Grant started writing about the mortgage industry as Special Reports Editor for National Mortgage News in 1997.

He subsequently served as managing editor for Origination News Magazine, Broker Magazine and Mortgage Technology magazine. Later, he served as editor for October Research Corp.'s Real Estate Technology Insight industry newsletter. He has written feature articles for Mortgage Banking and Scotsman Guide, served as a featured columnist for HousingWire and a blogger on real estate firm Zillow's consumer-facing blog.

Today, he serves as contributing editor for National Mortgage Professional, producer for Mortgage News Network and blogs at http://www.gkexaminer.com.

End Notes

[i] Why distributed ledger technology (DLT)? Per the U.S. Federal Reserve working paper, "The driving force behind efforts to develop and deploy DLT in payments, clearing, and settlement is an expectation that the technology could reduce or even eliminate operational and financial inefficiencies, or other frictions, that exist for current methods of storing, recording, and transferring digital assets throughout financial markets. The purported benefits of DLT that could address these frictions, including improved end-to-end settlement speed, data auditability, resilience, and cost efficiency, have led industry participants to investigate the application of DLT to a wide variety of PCS processes. Proponents of the technology have claimed that DLT could help foster a more efficient and safe payments system, and may even have the potential to fundamentally change the way in which PCS activities are conducted and the roles that financial institutions and infrastructures currently play." See "Distributed ledger technology in payments, clearing, and settlement", U.S. Federal Reserve, 2016-095.

[ii] Per Wikipedia, "The financial crisis of 2007–2008, also known as the global financial crisis and the 2008 financial crisis, is considered by many economists to have been the worst financial crisis since the Great Depression of the 1930s. It began in 2007 with a crisis in the subprime mortgage market in the U.S., and developed into a full-blown international banking crisis with the collapse of the investment bank Lehman Brothers on September 15, 2008." See https://en.wikipedia.org/wiki/Financial_crisis_of_2007%E2%80%932008.

[iii] Trade date + 0 – same day settlement (T+0). Per Investopedia, "Whenever you buy or sell a stock, bond, or mutual fund, there are two important dates of which you should always be aware: the transaction date and the settlement date. The abbreviations T+1, T+2, and T+3 refer to the settlement dates of security transactions which occur on a transaction date plus one day, plus two days, and plus three days, respectively. 'T' is the transaction date." See http://www.investopedia.com/ask/answers/03/061303.asp

[iv] Per Wikipedia, "The Art of War is an ancient Chinese military treatise dating from the 5th century BC and is attributed to the ancient Chinese military strategist Sun Tzu ("Master Sun", also spelled Sunzi). The text is composed of 13 chapters, each of which is devoted to one aspect of warfare and it is commonly thought of as a definitive work on military strategy and tactics." https://en.wikipedia.org/wiki/The_Art_of_War.

[v] This statement was made before an investment during the Great Recession. "Because of an astute investment made in Bank of America six years ago while the bank was struggling, Warren Buffett is about to make a quick $12 billion and become the financial institution's largest shareholder. Buffett's Berkshire Hathaway said Friday it will exercise warrants in Bank of America that will allow it to acquire 700 million common shares in the bank at an exercise price of just $7.14 each, or about $5 billion. Bank of America shares closed Thursday at $24.32 a share, making that stake actually worth more than $17 billion. So, it's a quick $12 billion paper profit for the Oracle of Omaha." CNBC.com, "Warren Buffett just made a quick $12 billion on a clever Bank of America investment", June 30, 2017, https://www.cnbc.com/2017/06/30/warren-buffett-just-made-a-quick-12-billion-on-bank-of-america.html

[vi] "Banks lost 4% of their total mortgage market share year over year, falling from 47% to 43%. Nonbank lenders had a small lead over banks last year with a 48% market share, which is up from 44% in 2014. Credit unions were unchanged at 9%." Bankrate.com, Quarterly Originations by Lender, October 19, 2016."

[vii] Per Wikipedia, "The Gramm–Leach–Bliley Act (GLBA), also known as the Financial Services Modernization Act of 1999. With the bipartisan passage of the Gramm–Leach–Bliley Act, commercial banks, investment banks, securities firms, and insurance companies were allowed to consolidate. Furthermore, it failed to give to the SEC or any other financial regulatory agency the authority to regulate large investment bank holding companies."
https://en.wikipedia.org/wiki/Gramm%E2%80%93Leach%E2%80%93Bliley_Act

[viii] It is worth noting that as homeowner rates hit 40-year lows, renters will pay "more than half of their incomes in rent in 2025 …", see "Renters Will Continue to Struggle for the Next Decade, Harvard Study Says", Dow Jones & Company, Kusisto, Laura, 2015.

[ix] "A Cheat Sheet on the De-globalization of the Financial World", Bloomberg.com, Verma, Sid, December 6, 2016.

[x] While not explicitly listed here, the use of machine learning would for our purposes fall under AI. Author Samuel (noted computer scientist and the person who popularized the term) defines machine learning as, "computers (with) the ability to learn without being explicitly programmed."

[xi] Love of Self (French phrase).

[xii] "New projections show that 14 million households will be formed by 2024…" "Fannie, Freddie Reshaping Mortgage Market without Congress' Help", National Mortgage News, Berry, Kate, October 26, 2015.

[xiii] J.P. Morgan Chase & Co., Bank of America Corp., Wells Fargo & Co., Citigroup Inc., and Goldman Sachs Group Inc.

[xiv] Some FSBO specialists will remember the days where we rallied the organizations to create enterprise initiatives (fill in the blank for the discipline, data, server, archiving, due diligence, et al) which generated great interest, set high expectations, produced lots of meetings and paper, and still fell short when it can to results.

[xv] Several FSBO's, in response to disintermediation and commoditization of their offerings, have embarked on a series of steps and strategies to compete in viable markets on their employee and outsourcer capabilities—when assembled across the financial supply chains (see Chapter 9).

[xvi] While there was a nearly 15% reduction of debt levels for households during the Great Recession (from 2007-2009), since 2010 debt levels have substantial climbed and now exceed pre-recessionary measurements not only in total outstanding, but also for measures to repay.

[xvii] "The U.S. Homeownership Rate Falls Again, Nearing a 48-Year Low", The Wall Street Journal, April 28, 2016.

[xviii] Economists have increasingly become pessimistic on the U.S. consumer's ability to deal with retirement and as debt levels rise, saving rates as measured by salaries have dramatically declined since 2015.

[xix] An example of this can be seen in a recent, and very detailed, consumer complaint filed by the CFPB. "The Consumer Financial Protection Bureau (CFPB) today sued one of the country's largest nonbank mortgage loan servicers, Ocwen Financial Corporation, and its subsidiaries for failing borrowers at every stage of the mortgage servicing process. The Bureau alleges that Ocwen's years of widespread errors, shortcuts, and runarounds cost some borrowers money and others their homes. Ocwen allegedly botched basic functions like sending accurate monthly statements, properly crediting payments, and handling taxes and insurance. Allegedly, Ocwen also illegally foreclosed on struggling borrowers, ignored customer complaints, and sold off the servicing rights to loans without fully disclosing the mistakes it made in borrowers' records. The Florida Attorney General took a similar action against Ocwen today in a separate lawsuit. Many state financial regulators are also independently issuing cease-and-desist and license revocation orders against Ocwen for escrow management and licensing issues today. "Ocwen has repeatedly made mistakes and taken shortcuts at every stage of the mortgage servicing process, costing some consumers money and others their homes," said CFPB Director Richard Cordray. "Borrowers have no say over who services their mortgage, so the Bureau will remain vigilant to ensure they get fair treatment.", See https://www.consumerfinance.gov/about-us/newsroom/cfpb-sues-ocwen-failing-borrowers-throughout-mortgage-servicing-process/

[xx] "Today, the Consumer Financial Protection Bureau (CFPB), authorities in 49 states, and the District of Columbia filed a proposed court order requiring the country's largest nonbank mortgage loan servicer, Ocwen Financial Corporation, and its subsidiary, Ocwen Loan Servicing, to provide $2 billion in principal reduction to underwater borrowers. The consent order addresses Ocwen's systemic misconduct at every stage of the mortgage servicing process. Ocwen must also refund $125 million to the nearly 185,000 borrowers who have already been foreclosed upon and it must adhere to significant new homeowner protections. Deceptions and shortcuts in mortgage servicing will not be tolerated," said CFPB Director Richard Cordray. "Ocwen took advantage of borrowers at every stage of the process. Today's action sends a clear message that we will be vigilant about making sure that consumers are treated with the respect, dignity, and fairness they deserve." See https://www.consumerfinance.gov/about-us/newsroom/cfpb-state-authorities-order-ocwen-to-provide-2-billion-in-relief-to-homeowners-for-servicing-wrongs/

[xxi] Numbers within categories vary by several million, but to give you an idea of the demographics within each mega-classification, "Millennials have surpassed Baby Boomers as the nation's largest living generation, according to population estimates released this month by the U.S. Census Bureau. Millennials, whom we define as those ages 18-34 in 2015, now number 75.4 million, surpassing the 74.9 million Baby Boomers (ages 51-69). And Generation X (ages 35-50 in 2015) is projected to pass the Boomers in population by 2028.", Pew Research, "Millennials overtake Baby Boomers as America's largest generation" April 25, 2016, http://www.pewresearch.org/fact-tank/2016/04/25/millennials-overtake-baby-boomers/

[xxii] "Here's a word of advice for companies in trouble: Don't make the public any angrier than necessary. That's the mistake Equifax Inc. repeated several times over in its careless handling of its careless loss of detailed identifying data on 143 million consumers, a

breach widely described as the worst in history. The company made a number of missteps, such as taking months to make the break-in public, and apparently running web server software with a known vulnerability. But the biggest question since the news broke has involved whether Equifax was trying to pull a fast one: Were worried consumers being forced to surrender their right to sue before they could find out if they were among the victims of the hack, or was that an urban myth?" See "Equifax Bungles the Details Over and Over Again", Bloomberg, September 11, 2017. https://www.bloomberg.com/view/articles/2017-09-11/equifax-bungles-the-details-over-and-over-again

[xxiii] See "How More Americans Are Getting a Perfect Credit Score", Bloomberg, August 16, 2017.

[xxiv] "America may finally be approaching what could arguably be called peak credit score. This year, the average national FICO number is 700, just above where it stood in October 2006, before the run-up to our most recent financial collapse. The ranks of 'super-prime' consumers—those with scores of 800 and up—have steadily increased since 2010, and now number over 41 million, more than consumers with scores of 600 or below." See "How More Americans Are Getting a Perfect Credit Score", Bloomberg, August 16, 2017.

[xxv] Verhage, Julie, "A Record Percentage of MBAs Don't Want to Work on Wall Street", Bloomberg, August 9, 2017, "A new survey shows that jobs outside of Wall Street are becoming increasingly enticing to recent MBA graduates. According to data from Training the Street, bulge bracket banks were a top employment choice for just 19 percent of respondents. Not only is that a drop of 7 percent from last year, it's also the lowest level in the eight-year history of the survey."

[xxvi] "This Could Be the Mortgage Industry's iPhone Moment", Matt Burns, November 24, 2015.

[xxvii] For a discussion on a model showing six major consumer shifts, see "Digital Disruptions: Six consumer trends and what businesses need to do now", Mckinsey, Duncan, Ewan, Hazan, Eric, and Roche, Kevin. March 2014.

[xxviii] For a larger discussion of how analytics are changing, see The Intelligent Distributor", Oliver Wyman, 2015 and Deloitte infographic, "6 steps to business analytics success".

[xxix] From Simply Psychology, "Maslow stated that people are motivated to achieve certain needs and that some needs take precedence over others. Our most basic need is for physical survival, and this will be the first thing that motivates our behavior. Once that level is fulfilled the next level up is what motivates us, and so on. This five-stage model can be divided into deficiency needs and growth needs. The first four levels are often referred to as deficiency needs (D-needs), and the top level is known as growth or being needs (B-needs).", https://www.simplypsychology.org/maslow.html.

[xxx] "Auto loans now account for 9.2% of all U.S. household debt." See graphic "Diver Debt", Bloomberg.

[xxxi] At the time of publication, Bloomberg news released that U.S. household debt exceeded $12.7 trillion—the highest in recorded history. "Income from wages pales in

comparison to debt loads." See "In Debt We Trust for U.S. Consumers", Bloomberg, August 10, 2017.

[xxxii] "ICOs have fueled a rapid ascent in the value of all cryptocurrencies, from about $17 billion at the start of the year - with bitcoin making up around 90 percent of that - to a record high close to $180 billion at the beginning of September, of which bitcoin represented less than half." See "Bitcoin slides on 'fraud' warning from JPMorgan's Dimon", Reuters, September 12, 2017, https://www.reuters.com/article/us-bitcoin-dimon/bitcoin-slides-on-fraud-warning-from-jpmorgans-dimon-idUSKCN1BO232

[xxxiii] "Third of global consumers open to Google, Amazon banking: survey", Technology News, January 11, 2017.

[xxxiv] Per Wikipedia, the Varian Rule holds that "A simple way to forecast the future is to look at what rich people have today; middle-income people will have something equivalent in 10 years, and poor people will have it in an additional decade. It is attributed to Google's chief economist Hal Varian. Andrew McAfee first called it 'the Varian Rule' in the Financial Times. An alternative interpretation put forth by The Guardian writer Evgeny Morozov is that 'Luxury is already here – it's just not very evenly distributed.' Some notable historical examples include anti-lock braking systems, electronic stability control and airbags which first appeared on high-end luxury vehicles before becoming commoditized on more mainstream automobiles."

[xxxv] Ten years after the Great Recession, FSBO revenue has gone from $300 billion to $230 billion and ROE from 18.5% to 6.7%. See "Banking's Lost Decade", Bloomberg, August 14, 2007.

[xxxvi] The accepted definition of an ICO is, "Initial coin offering (ICO) is an unregulated means of crowdfunding via use of cryptocurrency. The term is often confused with 'token sale' or crowd sale, which refers to a method of selling participation in an economy, giving investors access to the features of a particular project starting at a later date. ICOs, on the other hand, sell a right of ownership or royalties to a project." See Wikipedia and Investopedia.

[xxxvii] "Amazon is doing great damage to tax paying retailers. Towns, cities and states throughout the U.S. are being hurt - many jobs being lost!", President Trump tweeted Aug. 16, 2017. Amazon's stock fell in response. It should be noted as published in Politifact.com on the same day, "… Amazon's push to have Congress pass the Marketplace Fairness Act, which would effectively force all online retailers to abide by state sales tax, and said it makes sense from a business perspective for Amazon to want to level the playing field."

[xxxviii] As of this publication, and for future understanding of the assumptions made within this book, the latest from the Bureau of Labor Statistics states, "Total nonfarm payroll employment increased by 156,000 in August, and the unemployment rate was little changed at 4.4 percent, the U.S. Bureau of Labor Statistics reported today. Job gains occurred in manufacturing, construction, professional and technical services, health care, and mining. Household Survey Data. in August, the unemployment rate, at 4.4 percent, and the number of unemployed persons, at 7.1 million, were little changed. After declining earlier in the year, the unemployment rate has been either 4.3 or 4.4 percent since April." See https://www.bls.gov/news.release/empsit.nr0.htm

[xxxix] We believe that the U.S. Federal Reserve whitepaper on DLT (distributed edger technology) currently states the use of DLT within FSBO's the best, "The extent to which DLT will have an impact on the financial architecture may become clearer as the technology matures. The level of impact will depend at least in part on decisions about which components of DLT are adopted and how DLT arrangements are ultimately implemented. As the industry considers a wide variety of applications, DLT arrangements are likely to take on a range of forms, with different components of the technology being used for different purposes. Thus, this section reviews some of the key components of DLT that could be adopted in the context of payments, clearing, and settlement." See "Distributed ledger technology in payments, clearing, and settlement", U.S. Federal Reserve, 2016-095.

[xl] A decade after the Great Recession began, the courts are still settling the aftermath of failed institutions including the disposition of divisions belonging to Lehman Brothers. "Two affiliates of Lehman Brothers, the U.S. investment bank that collapsed in 2008 and fueled an economic crisis, filed for Chapter 11 bankruptcy late on Thursday, a reminder of the complexity of unwinding a global financial institution. he affiliates own residential mortgage-backed securities, real estate and stock in First Data Corp (FDC.N), which helps process credit card transactions, among other assets, according to papers filed in the U.S. bankruptcy court in Manhattan." See "Nine years on, another Lehman Brothers bankruptcy", Reuters, http://www.reuters.com/article/us-lehmanbros-bankruptcy/nine-years-on-another-lehman-brothers-bankruptcy-idUSKCN1BC5UP

[xli] "Globalization and technological change is here to stay. And the promise of those is very good - we know that it can raise standards of living," Federal Reserve Bank of Cleveland President Loretta Mester, per Reuters, "Globalization's castaways haunt central bankers", August 24, 2017.

[xlii] This is a liberty taken from Abraham Lincoln's speech, "A House Divided". Per Wikipedia, "The House Divided Speech was an address given by Abraham Lincoln, later President of the United States, on June 16, 1858 at what was then the Illinois State Capitol in Springfield, after he had accepted the Illinois Republican Party's nomination as that state's US senator." The exact quote is, "A house divided against itself cannot stand. I believe this government cannot endure, permanently, half slave and half free. I do not expect the Union to be dissolved — I do not expect the house to fall — but I do expect it will cease to be divided. It will become all one thing or all the other. Either the opponents of slavery will arrest the further spread of it, and place it where the public mind shall rest in the belief that it is in the course of ultimate extinction; or its advocates will push it forward, till it shall become lawful in all the States, old as well as new — North as well as South."

[xliii] The rise of Bitcoin has surged nearly 400% in the last year and there are now Chicago futures exchanges which will soon begin to take "bets" long or short on its movement. However, the CEO of JP Morgan Chase & Co on September 12, 2017, stated in article appearing in Bloomberg that, "he would fire any employee trading bitcoin for being 'stupid.' The cryptocurrency 'won't end well,' he told an investor conference in New York on Tuesday, predicting it will eventually blow up. 'It's a fraud' and 'worse than tulip bulbs.' If a JPMorgan trader began trading in bitcoin, he said, 'I'd fire them in a second. For two reasons: It's against our rules, and they're stupid. And both are dangerous.'"

[xliv] Banking executives have taken a "unique" view on cybercurrencies especially Bitcoin. Since any of the banks are coming together on their blockchain solutions via R3, the opposition to the early cybercurrency frontrunners has noticeably stepped up. For

example, Jamie Dimon, CEO of JP Morgan & Chase Co recently stated, "If you were in Venezuela or Ecuador or North Korea or a bunch of parts like that, or if you were a drug dealer, a murderer, stuff like that, you are better off doing it in bitcoin than U.S. dollars, so there may be a market for that, but it'd be a limited market." See Bloomberg, September 17, 2017, https://www.bloomberg.com/news/articles/2017-09-12/jpmorgan-s-ceo-says-he-d-fire-traders-who-bet-on-fraud-bitcoin

[xlv] At the time of publication, Bitcoin was approaching $5,000.

[xlvi] Frequently used by startups to raise funding, it is "A cross between crowdfunding and an initial public offering, ICOs involve the sale of virtual coins mostly based on the Ethereum blockchain, similar to the technology that underpins bitcoin. But unlike a traditional IPO in which buyers get shares, getting behind a startup's ICO nets virtual tokens -- like mini-cryptocurrencies -- unique to the issuing company or its network. That means they grow in value only if the startup's business or network proves viable, attracting more people and boosting liquidity." See "Bitcoin Tumbles as PBOC Declares Initial Coin Offerings Illegal", Reuters, September 4, 2017.

[xlvii] "Six new banks have joined a UBS-led (UBSG.S) effort to create a digital cash system that would allow financial markets to make payments and settle transactions quickly via blockchain technology. The group aims to launch the system late next year. Barclays (BARC.L), Credit Suisse (CSGN.S), Canadian Imperial Bank of Commerce (CM.TO), HSBC (HSBA.L), MUFG (8306.T) and State Street (STT.N) have joined the group developing the "utility settlement coin" (USC), a digital cash equivalent of each of the major currencies backed by central banks," See "Six big banks join blockchain digital cash settlement project", Reuters, August 31, 2017.

[xlviii] Per Wikipedia, "R3 (R3CEV LLC) is a distributed database technology company. It leads a consortium of more than 70 of the world's biggest financial institutions in research and development of blockchain database usage in the financial system. It is headquartered in New York City. It was founded in 2014 by David E Rutter."

[xlix] The idea of rapid innovation, can also be captured under the banner of technology experimentation. The idea that innovation can be expanded using experimentation has merit for cultures struggling with competitive factors and an opaque vision of "what's next". To use experimental procedures and processes does require discipline and rigor familiar with most scientific approaches to gather evidence and efficacy for customer approaches.

[l] Issued by the U.S. Federal Reserve on September 1, 2017, to ensure safe and soundness of the markets in times of crisis, new rules were implemented to stop the halting or breaking of contracts when a trading partner collapses (as happened repeatedly during the Great Recession). This allows regulators and the courts to act timely, but not be subject to millisecond market reactions. Per Janet Yellen, U.S. Federal Reserve Chairwoman, "The very largest banks are interconnected through substantial volumes of financial contracts, unless those contracts can be kept from unraveling, they can spark asset fire sales that can consume many firms." See "Fed Demands Wall Street Protections Against Lehman-Like Runs", Bloomberg, September 1, 2017.

[li] Commoditization surfaces when competitive products and services are functionally the same leading to a void of consumer enthusiasm and a perception that the provider's solution does not fit into the lifestyle of how financial products are delivered.

[lii] Consumers are in one model, estimated to work within nine taxonomies of digital assets and 29 distinct categories. See "Consumers are Tied to Many Digital Assets", Slideshare.net.

[liii] . https://www.aba.com/Tools/Infographics/Pages/Infographic-MillennialsBanking.aspx

[liv] Per Wikipedia, "The Keystone Cops (often spelled "Keystone Kops") were fictional incompetent policemen, featured in silent film comedies in the early 20th century. The movies were produced by Mack Sennett for his Keystone Film Company between 1912 and 1917."

[lv] Millennials are now the largest demographic within the U.S. They can be sub-segmented into younger (18-22), middle (23-29), and older millennials (30-38). See "Inside the Mind of Millennials: What Financial Marketers Need to Know", The Financial Brand, January 25, 2016.

[lvi] "U.S. homebuilding … of multi-family houses tumbled to a 10-month low, but strong job growth is expected to continue to support the housing market recovery. Housing starts declined 4.8 percent to a seasonally adjusted annual rate of 1.16 million units, hurt also by a drop-in groundbreaking on single-family projects." The housing industry has yet to regain growth-inspired building rates (i.e., 1.6 million units per year) a decade after widespread homeowner and consumer bankruptcies and foreclosures. See "U.S. housing starts fall as multi-family construction slumps", Reuters, August 16, 2017.

[lvii] "Be 'very afraid' about globalization's next phase", Money.cnn.com, Long, Heather, December 23, 2016.

[lviii] An extended discussion of what constitutes an "innovator" can be found in "Taking the measure of your innovation performance", Bain & Company. "Great innovators develop five critical capabilities—and always look at the process from both sides of the brain."

[lix] While the hope is banking costs will decline, "Costs as a proportion of revenue remain stubbornly highly". See "Banking's Lost Decade", Bloomberg August 14, 2017.

[lx] "The vulnerabilities within the financial system in the mid-2000s were numerous and, in hindsight, familiar from past financial panics. Financial institutions had assumed too much risk, especially related to the housing market, through mortgage lending standards that were far too lax and contributed to substantial overborrowing. Repeating a familiar pattern, the "madness of crowds" had contributed to a bubble, in which investors and households expected rapid appreciation in house prices. The long period of economic stability beginning in the 1980s had led to complacency about potential risks, and the buildup of risk was not widely recognized. As a result, market and supervisory discipline was lacking, and financial institutions were allowed to take on high levels of leverage. This leverage was facilitated by short-term wholesale borrowing, owing in part to market-based vehicles, such as money market mutual funds and asset-backed commercial paper programs that allowed the rapid expansion of liquidity transformation outside of the regulated depository sector. Finally, a self-reinforcing loop developed, in which all of the factors I have just cited intensified as investors sought ways to gain exposure to the rising prices of assets linked to housing and the financial sector. As a result, securitization and the development of complex derivatives products distributed risk across institutions in ways that

were opaque and ultimately destabilizing." Excerpt from U.S. Federal Reserve Chairwoman Janet Yellen, August 25, 2017, https://www.federalreserve.gov/newsevents/speech/yellen20170825a.htm

[lxi] It's a loose translation of a military strategy to create an army to act it is sometimes necessary to give them a reason to confront the challenges. By starting a "foot on fire", change is easier as it becomes a means to a greater solution or goal. Modern strategists sometimes refer to this same principle as a "burning platform". That is, if the platform you are on is on fire, you will either put out the fire and rebuild or flee for a different surrounding.

[lxii] Recent projections regarding reduced banking regulations would benefit the largest U.S. Banks the most—with them estimated to received $27 billion in pre-tax profits. See "Big U.S. Banks Could See Profit Jump 20% With Trump Deregulation", Bloomberg, August 23, 2017. https://www.bloomberg.com/news/articles/2017-08-23/big-u-s-banks-could-see-profit-jump-20-with-trump-deregulation

[lxiii] TBTF is a label that underneath represents a set of operating guidelines for FSBO's (banking and insurance) covering select dominant institutions and those with globally connectivity. These TBTF are viewed as systemically important especially in times of crisis or failure. If allowed to fail, these institutions could alter, harm or break the financial supply chains resulting in frozen markets, loss of assets or even as a prelude to a run on the domestic economic systems—a domino effect that if unchecked, would destroy the entire safeness and soundness of financial instruments.

[lxiv] As this book was going to print, the U.S. Federal Reserve issued new guidance for those institutions (approximately 30 of them) on the TBTF list. "The rule requires global systematically important banks (GSIBs) to amend the language in common financial contracts so they cannot be immediately canceled if the firm enters bankruptcy. By imposing new legal protections, regulators aim to prevent a run on a GSIB's subsidiaries that could be triggered if a large number of counterparties rush to terminate their contracts, as occurred in the case of Lehman Brothers in 2008." See "Federal Reserve finalizes rules to help unwind big banks", Reuters, September 1, 2017.

[lxv] Per Investopedia, "The Sarbanes-Oxley Act of 2002 (SOX) is an act passed by U.S. Congress in 2002 to protect investors from the possibility of fraudulent accounting activities by corporations. The SOX Act mandated strict reforms to improve financial disclosures from corporations and prevent accounting fraud. The SOX Act was created in response to accounting malpractice in the early 2000s, when public scandals such as Enron Corporation, Tyco International plc, and WorldCom shook investor confidence in financial statements and demanded an overhaul of regulatory standards." Read more: Sarbanes-Oxley Act Of 2002 - SOX http://www.investopedia.com/terms/s/sarbanesoxleyact.asp#ixzz4sQ4oznQs

[lxvi] In a common-sense realization of the impact of this rule, Federal Reserve (FED) Chairwoman, Janet Yellen stated on August 25, 2017, (and paraphrased by Bloomberg in the article "Yellen Distances Herself From Trump in Jackson Hole Speech"), "… some aspects of the Volcker Rule, which limits proprietary trading by banks, may be simplified while the Fed is taking steps to reduce 'unnecessary complexity' in rules affecting small banks."

[lxvii] "Cybersecurity is the body of technologies, processes and practices designed to protect networks, computers, programs and data from attack, damage or unauthorized access. In a computing context, security includes both cybersecurity and physical security." From, "What is Cybersecurity?", http://whatis.techtarget.com/definition/cybersecurity

[lxviii] There is an interesting document worth review, "The power to perform: Human capital 2020 and beyond", PwC, 2016.

[lxix] "Trump Cites Friends to Say Banks Aren't Making Loans. They Are.", Bloomberg, February 4, 2017, https://www.bloomberg.com/news/articles/2017-02-04/trump-cites-friends-to-say-banks-aren-t-making-loans-they-are

[lxx] The building of "new single-family homes give the economy a bigger boost than existing-home sales or the construction of new apartments, which is booming as more people rent…a single-family home supports three full-time jobs for a year in construction and ancillary services. In comparison, it estimates that construction of a condo or apartment unit supports one full-time job." "Why a Stronger Housing Sector Isn't Boosting the U.S. Economy That Much", WSJ, September 9, 2015.

[lxxi] It's worth noting the significant trend in owning a home and the downstream consequences of declining rates across the U.S. Economy. One of the key reasons home loans have declined is the supply of housing available—down to 4.2 months of supply from a healthy standard of 6 months. Also, affordability indexes for most major cities have reached pre-Great Recession levels shutting out many first-time homebuyers. Annual units sold, have stumbled from over 5.5 million annually to around 4.8 million single family homes.

[lxxii] "The (U.S.) Federal Reserve is committed to evaluating where reforms are working and where improvements are needed to most efficiently maintain a resilient financial system," Janet Yellen, Federal Reserve Chairwoman, August 25, 2017.

[lxxiii] It's interesting to note, especially giving the recent debates on racial extremism, that the government debt skyrocketed starting back in the Civil War after having been effectively wiped out from the post-Revolutionary War. As a percentage of the U.S. GDP, current levels have not been this high since World War II and show no signs of being balanced between fiscal policy and monetary policy. See https://en.wikipedia.org/wiki/History_of_the_United_States_public_debt

[lxxiv] "The threat to the stock and corporate bond markets underscores the bind that the Fed is in as it tries to scale down the extraordinary stimulus it gave the economy during and after the financial crisis. Its quantitative easing program has more than quadrupled its debt holdings, and Fed policy makers plan to start slimming down those investments soon now that the economy is on more solid ground. But cutting back could hurt markets and perhaps the broader economy in ways that are hard to forecast." See "Fed's Big Bond Unwind May Clobber U.S. Stocks, Corporate Debt", Bloomberg, August 21, 2017.

[lxxv] One of the goals explicitly stated by the U.S. Federal Reserve is to create and sustain a financial system able to withstand market shock of varying degrees of impact. As markets become increasingly global and currency virtual, new measures and tests are being adopted in a hope to stave off unforeseen disruptions.

[lxxvi] For a complete copy of U.S. Federal Reserve Chairwoman, Janet Yellon's speech at Jackson Hole financial symposium, see https://www.federalreserve.gov/newsevents/speech/yellen20170825a.htm

[lxxvii] "It will be of little avail to the people, that the laws are made by men of their own choice, if the laws be so voluminous that they cannot be read, or so incoherent that they cannot be understood; if they be repealed or revised before they are promulgated, or undergo such incessant changes that no man, who knows what the law is to-day, can guess what it will be to-morrow." Alexander Hamilton, 1788, deemed by some as "The Father of Modern Banking" due to his modeling of the financial system after Great Britain.

[lxxviii] "71% of all B2B research starts with unbranded searches. Previous research showed that those involved in the B2B buying process are already 57% on the way down the purchase path before they'll actually perform an action on a brand's website.", see "What You Thought About B2B Marketing and Millennials Might Be Wrong!", Linkedin.com, Michael Brio, March 17, 2015.

[lxxix] The trend for current home markets across the U.S., show that the time needed to sell a home has dropped below 30 days—a 20% decline in the cycle in just a single year. Additionally, of the homes that sell, over one-half sell within a single month after having been listed.

[lxxx] For an older, but in-depth review of what Toronto Dominion Bank values, see "Creating a Competitive Advantage", TDBank.com.

[lxxxi] "The soft underbelly of companies outside the banking sector was exposed again this month when hackers leaked details of nearly 37 million clients… More than half the value of companies worldwide is in intangible assets, such as intellectual property, much of which is stored on computers and could therefore be vulnerable to hackers." See "Investors still in the dark as cyber threat grows", Reuters, August 28, 2015, http://uk.reuters.com/article/uk-funds-cybersecurity-idUKKCN0QX1M420150828

[lxxxii] See Bloomberg, June 11, 2015, https://www.bloomberg.com/news/special-reports/what-is-code

[lxxxiii] "China, Russia, soon all countries w strong computer science. Competition for AI superiority at national level most likely cause of WW3 imo." For a tweet posted by Elon Musk, CEO Telsa and SpaceX, see Market Watch, September 5, 2017, http://www.marketwatch.com/story/elon-musk-predicts-world-war-iiibut-not-for-the-reason-you-might-guess-2017-09-05

[lxxxiv] There are countless historical examples, but there is one that standouts out. The transistor was invested over 50 years before it found its way into consumer electronics and laid the foundation for the advanced microprocessors of today. The transistor was coveted for military usage until Sony applied it to a portable consumer electronic device—a Walkman was born. While ancient by today's standards, this invention was only relevant for modern society after it found its "Innovative Relevance". And the rest the say, is history.

[lxxxv] It is for valid reasons that FSBO's (including the credit agencies) vastly increase their capabilities and staffing in the area of cybersecurity. The latest event was executed against, "Equifax Inc. said its systems were struck by a cybersecurity incident that may have affected about 143 million U.S. consumers, shedding light on what could be

ranked as one of the largest breaches in history. Intruders accessed names, Social Security numbers, birth dates, addresses and driver's license numbers…". See "Equifax Says Cyber Attack May Have Hit 143 Million Customers", Bloomberg, September 7, 2017.

[lxxxvi] "Wal-Mart and Google Partner to Challenge Amazon", Wall Street Journal online, August 23, 2017, https://www.wsj.com/articles/wal-mart-and-google-partner-to-challenge-amazon-1503460861

[lxxxvii] "The amount of money banks lend relative to their deposits has also declined." This may be attributable to recapitalization levels, risk aversion, availability of investment or regulatory compliance. See "Banking's Lost Decade", Bloomberg, August 14, 2017.

[lxxxviii] Embracing the digital revolution is about bold management, not just technology. See "Leading a 36-degree Digital Transition", Bain & Company.

[lxxxix] "How companies get stuck throwing good money after bad ideas—and what that mistake is costing them", see "The Innovation Death Spiral", Accenture, 2011.

[xc] For a discussion on board level response to crisis, see "Crisis Management, Focus on Building Crisis-Ready Boards", Deloitte, May 2015.

[xci] If you are seeking a discussion on what the impacts for EC's could potentially touch, as the world's consumers begins a multi-decade phase out of physical instruments, then we might suggest additional readings from the Federal Reserve's Virtual Currency Debriefs, The Age of Crypto Currency by Vigna and Casey, or even the open-source efforts of the Linux Foundation to Advance Blockchain Technology. The quantum change of cryptocurrencies requires a comprehensive shift in technology, procedures, clearing, settlements, regulations, and yes, marketing.

[xcii] CMO's are increasingly digitally driven and market "savvy". For a statistical overview of how CMO's view their customers and how they are changing market approaches, see "Data-Driven and Digitally Savvy", Forbes Infographic, www.e.turn.com/sixthsense.

[xciii] "Carmakers face electric reality as combustion engine outlook dims", Reuters, September 12, 2017, http://www.reuters.com/article/us-autoshow-frankfurt-electrics/carmakers-face-electric-reality-as-combustion-engine-outlook-dims-idUSKCN1BN00X

[xciv] "Vikram Pandit, who ran Citigroup Inc. during the financial crisis, said developments in technology could see some 30 percent of banking jobs disappearing in the next five years." See "Pandit Says 30% of Bank Jobs May Disappear in Next Five Years", Bloomberg, September 13, 2017.

[xcv] It is worth noting that U.S. debt, since 2008, has risen approximately 100% from approximately $10 trillion to $20 trillion. U.S. Government Publishing Office, U.S. Treasury. See Bloomberg, "Trump Shutdown Threat Complicates Congress's Debt Ceiling Plans," August 23, 2017. https://www.bloomberg.com/news/articles/2017-08-23/trump-shutdown-threat-complicates-congress-s-debt-ceiling-plans

[xcvi] "Wall Street's biggest firms are using technologies including machine learning and cloud computing to automate their operations, forcing many employees to

adapt or find new positions. 'Everything that happens with artificial intelligence, robotics and natural language -- all of that is going to make processes easier,' said Pandit, who was Citigroup's chief executive officer from 2007 to 2012. It's going to change the back office.'" See "Pandit Says 30% of Bank Jobs May Disappear in Next Five Years", Bloomberg, September 13, 2017. https://www.bloomberg.com/news/articles/2017-09-13/ex-citi-ceo-pandit-says-30-of-bank-jobs-at-risk-from-technology

[xcvii] "Who will Own Mortgage Assets?", MBA.org, Fratantoni, Michael, 2014.

[xcviii] "Changing the strategy of a large bank, or any large company for that matter, is a bit like turning a supertanker. The momentum of the institution is so strong that the ability to change direction quickly is limited. After all, the … is on near-term operating decisions …" See Managing the Strategy Journey, McKinsey Quarterly, July 2012.

[xcix] "Building competitive advantage with software through a continuous delivery process", see "Continuous Delivery: A maturity Assessment Model", Forrester Consulting, March 2013.

[c] "Why Nonbank Lenders are the Future of Mortgages", National Mortgage News, Walen Christopher, February 16, 2016.

[ci] Attributable to Thomas Jefferson, but adapted for this discussion.

[cii] "The global digital ecosystem is emerging as a network of data and analytic services, data sources, and devices. Within the digital ecosystem, data flows across a variety of applications in cycles as fast as real-time." For a unique vendor discussion of digital transformation, see "Controlled and Managed Transformation, Unlocking Innovation with Data and Analytics", HCLTech.com and Gartner, June 2016.

[ciii] "When we look back to the mid-90s ad over the last 15 years, we saw lots of banks doing a me-too strategy, trying to be everything to everybody, changing their DNAs, doing things they weren't equipped to do. That was the anomaly." See "Beyond Banking: under attack on all sides", Financial Times, November 10, 2015.

[civ] In an acknowledgement that financial systems go through cycles, U.S. Federal Reserve Chairwoman Janet Yellen on August 25, 2017 stated from the annual Jackson Hole economic symposium, "I expect that the evolution of the financial system in response to global economic forces, technology, and, yes, regulation will result sooner or later in the all-too-familiar risks of excessive optimism, leverage, and maturity transformation reemerging in new ways that require policy responses,"

Made in United States
North Haven, CT
10 November 2026

22095611R00196